Stan Schuman, MD, DrPH

Rainbows in Washtubs
Diagnostic Mysteri
in Agromedicine

Pre-publication
REVIEWS,
COMMENTARIES,
EVALUATIONS . . .

"A delightful read. Dr. Schuman's small book is a quick read, but the ideas in it are anything but small. Dr. Schuman came to South Carolina at the invitation of Dr. Hiram B. Curry, founder of the Department of Family Medicine of the Medical University of SC. Dr. Curry was fond of saying, 'Some doctors have one year of experience twenty-five times, some have twenty-five years of experience—they are continually learning.' Schuman is obviously always learning. His insights are valuable for every person who seeks to understand and improve his or her surroundings, not just physicians. Written like a conversation between two friends, the 'mysteries' draw you in and make you think."

William M. Simpson Jr., MD
Professor of Family Medicine,
Medical Director, South Carolina Agromedicine Program,
Medical University of South Carolina

"Dr. Schuman has penned an insightful book that will appeal to physicians, researchers, agricultural extension workers, farmers, and anyone with an interest in the myriad of relationships between human health and the field of agriculture. Each short chapter is a case study of a different medical challenge. Dr. Schuman's journey extends from the simplest of farm situations to the sophisticated halls of medical conferences, but his direct and pragmatic style of writing prevails throughout. Teachers will love the book. Every case is a real-life example of the successes, as well as the shortcomings, of modern medicine. My copy of the book already has many pages flagged that I will be using in lectures and seminars for years to come."

James W. Rushing, PhD
Professor and Center Director,
Clemson University,
Coastal Research and Education Center,
Charleston, South Carolina

More pre-publication
REVIEWS, COMMENTARIES, EVALUATIONS . . .

"Absorbing. . . . He does it with wit, grace, keen intellect, and a selfless admiration for people he serves and with whom he works. Professional workers and students in any of the relevant fields will find guidance on scientific method based on hard evidence and an inquiring mind looking for and integrating information from multiple sources. Every reader will be impressed with methods of problem solving that are strengthened by patience, tolerance, and respect for people and their concerns, no matter how far outside usual experience and practice."

C. Arden Miller, MD
Professor Emeritus, Maternal and Child Health, School of Public Health, University of North Carolina; Past President, American Public Health Association

"This book is an engaging, enlightening view into the dynamic interactions of health care professionals with individuals involved in diverse agricultural pursuits. Both parties face diverse medical issues including food safety, the integrity of the environment, farm safety, and rural health needs and preventative medicine in the face of rapidly evolving technologies. Agromedicine is part of the discipline of occupational and environmental medicine. Although ancient in origin, it remains mysterious to many in modern society, including health care professionals.

We are treated to glimpses of Dr. Schuman's fascinating personal history as an investigator for the U.S. Public Health Service in Mississippi, a general pediatrician in St. Louis, an epidemiologist at the University of Michigan, and as the founder of the Agromedicine Program of Clemson University and the Medical College of South Carolina where he is Emeritus Professor of Family Medicine. The vignettes show a clarity of message that is essential to elucidating medical mysteries. In addition, Dr. Schuman cleverly shows that hubris has no role in solving agromedical problems, where complex societal and economic barriers are often involved. These stories will be entertaining and informative to expert medical practitioners and the general public. Importantly, Dr. Schuman's detailed medical mysteries also serve to elucidate the fact that medicine is frequently an 'art' rather than pure 'science.'"

Joshua D. Nosanchuk, MD
Assistant Professor, Division of Infectious Diseases, Departments of Medicine and Microbiology and Immunology, Jack and Pearl Resnick Campus, Albert Einstein College of Medicine at Yeshiva University

More pre-publication
REVIEWS, COMMENTARIES, EVALUATIONS . . .

"In *Rainbows and Washtubs* Dr. Stan Schuman, a respected professor, clinician, and epidemiologist, addresses a crucial but little-known field of practice: agromedicine, i.e., occupational and environmental medicine as it applies to exposures in agriculture, animal husbandry, and related endeavors. He engages the reader with short, often tart, tales of tracking down the source of mysterious illnesses. He approaches each case with ingenuity and perseverance, as well as with his wide knowledge of environmental medicine. Unlike many doctors today, Dr. Schuman also pays attention to every detail. From Peruvian lilies to the blow flies, from fire ants to raspberries, from contact dermatitis to hog farmer's pulmonary disease, the reader encounters an array of fascinating characters and cases, presented by a fine storyteller. This is an interesting introduction to agromedicine and a welcome addition to medical mystery genre pioneered so successfully by Berton Rouche."

Jack Coulehan, MD, MPH
Professor and Division Head,
Medicine in Society,
Stony Brook University

"This is the colorful and arresting story of Dr. Stan Schuman's remarkable career and his eventual devotion to the practice of agromedicine in rural South Carolina. Drawn mostly from personal experience, it is told in brief accounts of people and communities faced with health threats: pesticide exposures, insect infestations, allergic reactions to plants, infections, environmental hazards, etc. Medical facts are presented in combination with vivid descriptions of interactions and conversations among patients, community members, and medical colleagues. In these descriptions of agromedicine, Dr. Schuman repeatedly stresses the importance of epidemiologic observations. Beyond agromedicine and epidemiology, the book is also strongly directed at the importance of teaching. By combining several professional themes (medical practice, epidemiology, public health), the book provides a useful, concrete text for teaching students in medical schools and postgraduate departments."

Clark W. Health Jr., MD
Medical Epidemiologist, Retired;
CDC Employee 1960-1962, 1965-1985;
American Cancer Society Employee,
1988-1998

HMP

The Haworth Medical Press®
An Imprint of The Haworth Press, Inc.
New York

NOTES FOR PROFESSIONAL LIBRARIANS AND LIBRARY USERS

This is an original book title published by The Haworth Medical Press®, an imprint of The Haworth Press, Inc. Unless otherwise noted in specific chapters with attribution, materials in this book have not been previously published elsewhere in any format or language.

CONSERVATION AND PRESERVATION NOTES

All books published by The Haworth Press, Inc., and its imprints are printed on certified pH neutral, acid-free book grade paper. This paper meets the minimum requirements of American National Standard for Information Sciences-Permanence of Paper for Printed Material, ANSI Z39.48-1984.

DIGITAL OBJECT IDENTIFIER (DOI) LINKING

The Haworth Press is participating in reference linking for elements of our original books. (For more information on reference linking initiatives, please consult the CrossRef Web site at www.crossref.org.) When citing an element of this book such as a chapter, include the element's Digital Object Identifier (DOI) as the last item of the reference. A Digital Object Identifier is a persistent, authoritative, and unique identifier that a publisher assigns to each element of a book. Because of its persistence, DOIs will enable The Haworth Press and other publishers to link to the element referenced, and the link will not break over time. This will be a great resource in scholarly research.

Rainbows in Washtubs

Diagnostic Mysteries in Agromedicine

THE HAWORTH MEDICAL PRESS®
Titles of Related Interest

Psychotherapy and the Somatizing Patient edited by E. Mark Stern and Virginia Fraser Stern

Fibromyalgia, Chronic Fatigue Syndrome, and Repetitive Strain Injury: Current Concepts in Diagnosis, Management, Disability, and Health Economics edited by Andrew Chalmers, Geoffrey Owen Littlejohn, Irving E. Salit, and Frederick Wolfe

Lost Voices: Women, Chronic Pain, and Abuse by Nellie A. Radomsky

Disease, Pain, and Suicidal Behavior by Elsebeth Stenager and Egon Stenager

Autogenic Training: A Mind-Body Approach to the Treatment of Fibromyalgia and Chronic Pain Syndrome by Micah R. Sadigh

Minding the Body: Psychotherapy in Cases of Chronic and Life-Threatening Illness edited by Ellyn Kaschak

The Concise Encyclopedia of Fibromyalgia and Myofascial Pain by Roberto Patarca-Montero

The Psychopathology of Functional Somatic Syndromes: Neurobiology and Illness Behavior in Chronic Fatigue Syndrome, Fibromyalgic, Gulf War Illness, Irritable Bowel, and Premenstrual Dysphoria by Peter Manu

Rainbows in Washtubs
Diagnostic Mysteries in Agromedicine

Stan Schuman, MD, DrPH

The Haworth Medical Press®
An Imprint of The Haworth Press, Inc.
New York

For more information on this book or to order, visit
http://www.haworthpress.com/store/product.asp?sku=5851

or call 1-800-HAWORTH (800-429-6784) in the United States and Canada
or (607) 722-5857 outside the United States and Canada

or contact orders@HaworthPress.com

Published by

The Haworth Medical Press®, an imprint of The Haworth Press, Inc., 10 Alice Street, Binghamton, NY 13904–1580.

PUBLISHER'S NOTE
The development, preparation, and publication of this work has been undertaken with great care. However, the Publisher, employees, editors, and agents of The Haworth Press are not responsible for any errors contained herein or for consequences that may ensue from use of materials or information contained in this work. The Haworth Press is committed to the dissemination of ideas and information according to the highest standards of intellectual freedom and the free exchange of ideas. Statements made and opinions expressed in this publication do not necessarily reflect the views of the Publisher, Directors, management, or staff of The Haworth Press, Inc., or an endorsement by them.

This book has been published solely for educational purposes and is not intended to substitute for the medical advice of a treating physician. Medicine is an ever-changing science. As new research and clinical experience broaden our knowledge, changes in treatment may be required. While many potential treatment options are made herein, some or all of the options may not be applicable to a particular individual. Therefore, the author, editor and publisher do not accept responsibility in the event of negative consequences incurred as a result of the information presented in this book. We do not claim that this information is necessarily accurate by the rigid scientific and regulatory standards applied for medical treatment. **No warranty, express or implied, is furnished with respect to the material contained in this book. The reader is urged to consult with his/her personal physician with respect to the treatment of any medical condition.**

Identities and circumstances of most individuals discussed in this book have been changed to protect confidentiality.

Cover design by Marylouise E. Doyle.

Library of Congress Cataloging-in-Publication Data

Schuman, Stanley H., 1925-
Rainbows in washtubs : diagnostic mysteries in agromedicine / Stanley H. Schuman.
p. cm.
ISBN: 978-0-7890-3276-8 (hard : alk. paper)
ISBN: 978-0-7890-3277-5 (soft : alk. paper)
1. Agriculture—Health aspects—Case studies. 2. Agricultural laborers—Health and hygiene—Case studies. 3. Medicine, Rural—Case studies. I. Title.
RC965.A5S34 2007
610.2'463—dc22

2007000544

To the readers and writers of medical suspense narratives,
but mostly, to the patients who are central to each story.

And to the girl of my dreams . . .
past, present, and future . . .
Joanna

ABOUT THE AUTHOR

Stanley H. Schuman, MD, DrPH, LLD, is Professor Emeritus in the Department of Family Medicine at the Medical University of South Carolina, Charleston.

CONTENTS

Author's Note

While all incidents described in the narratives are true, as best as I can recall them, most places and names have been given pseudonyms to protect confidentiality.

DEFINITION

The term *agromedicine* can be defined as a field of endeavor blending the skills of medicine and agriculture to promote the health of farm workers, farm families, and consumers of farm products.

The following provide examples of the activities in this field:

Journal of Agromedicine
North American Agromedicine Consortium
Agromedicine Program of the University of Clemson and MUSC in South Carolina

ABBREVIATIONS

In addition, many abbreviations are used throughout the book. Some are commonly known but others are not. Some are defined in text but others are not. A short list of these abbreviations is provided here for your reference.

Rainbows in Washtubs

doi:10.1300/5851_a

AID	Agency for International Development
CDC	Centers for Disease Control and Prevention
EPA	Environmental Protection Agency
FAO	Farm and Agriculture Organization
ILO	International Labor Organization
MUSC	Medical University of South Carolina
NCI	National Cancer Institue
NIH	National Institutes of Health
OEM	occupational and environmental medicine
PHS or **USPHS**	Public Health Service or United States Public Health Service
RIFA	red imported fire ant
UN	United Nations
WHO	World Health Organization

Chapter 1

Finding Rainbows in Washtubs: An Unlikely Story

At times it's like walking in the footsteps of Joseph Goldberger, MD, USPHS, in the 1920s. I imagine that I'm watching him trying to track down the cause of an epidemic in poor Southerners with the butterfly rash on their cheeks, debilitating diarrhea, unexplained dementia and weight loss. How did he manage to keep studying pellagra, a disease that baffled all of the medical, epidemiologic, and infectious disease experts in the United States and in endemic areas of the disease in Europe?

For years in the field and in the lab, he kept observing the different settings where pellagra was most severe and most hopeless: mental institutions, orphanages, destitute textile mill towns, and penitentiaries. Depressing, at best, as one theory of causation after another failed to explain the cure or the prevention of the disease.

Finally, against all odds and despite sharp criticism by his superior officers in the PHS Corps, he targeted the telltale diet of poor inmates and orphans (corn mush and molasses). Goldberger believed that monotonous diet was the sole cause of the mysterious malady, not an infectious agent.

It was almost a decade later that biochemists could finally isolate niacin and support Goldberger's thesis that some vital nutrient was missing from the meager diet of impoverished victims of clas-

Rainbows in Washtubs

doi:10.1300/5851_01

sical pellagra. Only a balanced diet of fresh fruits and vegetables and quality proteins provides the essential vitamin.

Soon, a succession of discoveries and chemicals would provide the foundation of the new science of nutrition. Vitamin A earned several chemists the Nobel Prize in Physiology or Medicine.

At times, reading of Goldberger's character and devotion to medical detective work, I feel a kind of kinship with Joe, especially this hot, humid afternoon in one of the poorest agricultural counties in our state. I find myself passing time, brushing gnats, seeking a little shade—no hint of a breeze—outside this abandoned sharecropper's shack. Inside, my chemist is swabbing baseboards, the baby's crib, and the dirt-infested corners of two concrete slab rooms. Maybe there will be a lingering trace of household contamination. Here I stand, waiting, beside our ice chest cooler in a weedy, dusty yard, for the next rack of labeled specimens. The sun so brutal for farmhands, so essential for rows of tender cotton plants that reach right up to the concrete-slab cabin. Cotton is the major cash crop here, unless any one of the hazards prevail. Usually it takes dozens of applications of agrichemicals in a single season to control weeds, bollworms, thrip, nematodes, rust, etc.

Sweat dripping, I wait in muggy heat, gnats busy and buzzing. The air-conditioned lab at the medical center seems hours and miles away. The center, where, if I had more sense, I'd still be making white-coat rounds with docile interns, role modeling tertiary care. What am I doing here of all places, shooing a stray dog rifling piles of garbage strewn all about, watching scrawny chickens scratching gravel in the dust? Here simply because our moonlighting resident MD phoned me in the middle of the night, reporting his quandary and asking me to come and investigate. The family doctor agreed.

Dr. Anderson was also curious about this poor tenant farmer's family getting into so much sudden sickness; three children and both parents filled the tiny ER Friday and Saturday with acute signs of some kind of viral infection. At first, the infant appears, desperate with pneumonia, and is rushed off to the children's hospital.

Then the six-year-old arrives with a stiff neck, ataxia, and muscle twitching (negative lumbar puncture). Next the nine-year-old comes in with severe cramps and diarrhea. Finally, "both parents not feelin' good" exhibit cramps and blurred vision.

After a while Dr. Anderson and the resident sort out the cases (confused at first by the patients' presenting themselves to the ER at different times with different last names). The lack of fever pointed to some type of household poisoning. A trial atropine and 2-Pam treatment seemed to relieve some of the symptoms and suggested a form of organophosphate pesticide toxicity.

But now, the question of the common source of the toxic agent kept tugging at the rural doctor's mind. Airborne, waterborne, or foodborne? Household drift from treated fields? Intentional homicide, making it a case for the sheriff? Report to the remote health department? Or consult with us at the university hospital? Or, more likely as not, forget it.

With or without investigation, life in this hard-scrabble corner of the universe will go on as usual. With few skills and less education, this household will toil to eat, pray at the Baptist church for safer passage to heaven, and hope to recover soon to return to the fields that sustain them. Standing by the ice chest, I figure the chances of finding anything from week-old samples on the mass spectrograph are next to nil. Just another unexplained, nearly fatal incident, an understandably delayed diagnosis, and bad luck. *So be it,* I admit, brushing at more gnats.

But then, under a scrawny oak with a tire-swing, my eyes focus on a huge iron wash tub, filled to the brim. Its surface sparkles brilliantly in the sunlight. I move closer for a good look. Half-floating, half-submerged in the murky tub are several wool Army blankets, dark and dank. Another cloud passes overhead, letting a second shaft of sunlight hit the grimy water—wow! I can't believe it!

An iridescent oil ring was floating on the surface, and another, and another! Now I suspect, no, now I *know* this is no dry run. The specimens will come out positive. Most likely it is oil-soluble concentrated organophosphate, fitting the clinical picture to a *T.* The common source settled beyond a doubt. Field study over.

. . . Three ways to absorb the toxin: inhalation, skin absorption, and ingestion! Army blankets saturated with chemicals against bedbugs and roaches on a rampage. Every food surface a chemical battlefield. The baby's crib and mattress soaked with toxin . . .

What next? The long ride home. The sadness of twentieth century rural poverty, ignorance, and neglect. Household misuse of an outdoor agrichemical. Residents against vermin—an old story. A mother's frantic effort to protect her baby from roaches in the crib. The good neighbor offering her the best sure-kill chemical—so much better than drugstore stuff.

What next after this Goldberger moment? I know Doctor Anderson will not be surprised by the workup. He will recall occupational mixer-loader cases from the distant past, teaching the resident. Sadly, and devotedly, he will attend the Baptist church Sunday to help the minister explain to the congregants again:

> "Never borrow a farm chemical for roaches, rats, or ants!
> Never treat baby's crib, the kitchen, or blankets!
> E'nuff said!
> Amen!"

Chapter 2

Client-Based Research: Tools That Work

Problem solving in agromedicine is little different from in the practice of clinical medicine, or investigational epidemiology. They all start with formulating the questions: Why? Why not? How come? So what?

The all important initiative for the investigator begins with the presenting question of the patient or the client (farmer, farm family, consumer, extension agent).

Typical Starting Point Questions

Typical human health related questions have a way of crossing usual accepted disciplines and specialties. The all-purpose general practitioner of medicine or agriculture has given way to modern specialists who may be hard to reach. The physician may feel uncomfortable with agriculture-related questions, and the land-grant university expert may feel uncomfortable with medical questions.

Here are some typical interdisciplinary agromedical questions:

- Why do I get this dermatitis (recurring severe rash) when I do floral arrangements?
- Why do I break out in hives when I am stung by fire ants?

Rainbows in Washtubs

doi:10.1300/5851_02

- Is it safe to drink well water next to the peach farm that uses agrichemicals?
- Should I feed my baby only organically grown fruits and veggies?
- Why am I short of breath in my dad's indoor hog operation?
- Is it safe to resume poultry processing after my kidney transplant?
- Is my wife's breast cancer due to her helping out on the farm (organochlorines)?
- Is it safe to give Alar-treated apples/applesauce to my children?
- Why use hearing protection?
- Should I pay to fix my tractor with a roll-over bar?
- How do I react to attached ticks when they are swollen with blood?
- Why doesn't my spider bite heal?
- Why can't the doctor treat the parasites crawling "under my skin" at night, and why won't they go away?
- Could our fertility problem be caused by working with chemicals on the farm?
- Is it okay to eat the imported gourmet cheese from our friends in France even though it's not pasteurized?

None of these are earth-shaking questions. None are likely to require high technological resources of a major medical center or land-grant university. Yet to the concerned, hurting, confused, fearful client, the possible answers represent a form of healing through better understanding and, most important, clues to prevention of recurrences caused by re-exposure to agents in the environment. Assuming the patient/client has exhausted all of his or her usual sources of help and information (including neighbors, relatives, Web sites, radio talk shows, TV ads, etc.), any of these questions could confront and challenge the average medical or agriculture or public health professional.

Depending on the occupational setting of the agromedical resource person, he or she may politely dodge the client's query ("out of my field"). Some refer the caller to some other source of help (often a state or federal agency) or dismiss the question entirely.

The Essential Tools of the Problem Solver in Agromedicine

Caring/empathy
Competence/consultation/humility
Continuity/patience/follow-up
Curiosity/observation/openness
Commitment/perseverance
Chance/opportunity/luck/serendipity
Intuition/experience/patterning
Ability to research (data, epidemiological survey, clinical series, etc.)
Tolerance for failure/partial success/partial failure
Funding of the program/public service

Applying the Tools in Agromedicine

The narrative case studies in this book reveal a variety of presenting problems to be solved, or at least ameliorated, by using skills available at a medical university and agricultural university working as a public-service, community-based team. For example, the first five tools in the previous list are basic features of a skilled and dedicated primary care medical professional. In a parallel way, the skilled and dedicated agricultural researcher and extension agent is on call in his community of farmers and farm families for problem solving in a large variety of needs and concerns.

The next three tools are the working requirements of the scientific method, whether in basic biology, clinical medicine, agricultural research, or epidemiological investigation. The last two tools are defined by experience, which is the best teacher. There are limits to institutional support and the sustainability of any agromedical program.

Motivation and commitment help the individual and the team to endure inevitable setbacks in helping, alleviating, or preventing agromedical problems. This comes down to the ability to accept the odds for success, partial success, or partial or utter failure. Outcomes tend to average out between the extremes of fulfillment and disappointment.

Case Studies That Test the Mettle

Far from being cut and dried, responding to the variety and unpredictability of agromedical calls for help can be a challenging, stimulating, and rewarding experience, especially for those who enjoy interdisciplinary, investigational pursuits. Exhibit 2.1 illustrates a few of the cases requiring investigational skills and cross-disciplinary, medical-epidemiological methodology.

Another case example of agromedical diagnosis involves a series of doctor-patient visits at a university family medicine center. The patient was a middle-aged, retired war correspondent, one of the few women to achieve recognition in the male-dominated field. The clinician, Dr. Brown, was a top-notch diagnostician, admired by colleagues and students alike. The patient, Ms. Evans, presented in great distress, was angry at her landlord for repeatedly treating her duplex apartment with pest-control applications (termiticides) with offensive chemical odors that lingered for days afterward. Although she was not a susceptible asthmatic patient and had no immediate signs or symptoms of classic organophosphate poisoning, she was convinced (and was persuading Dr. Brown) that her health was suffering chemical-toxic impairment

EXHIBIT 2.1.
Three Client-Based Investigations in Agromedicine

I

Client/Patient: Young farmer with recent kidney transplant wishes to work in a proposed poultry-raising and shipping plant
Question: Are new poultry plant exposures safe for him or not?
Methodology:
A. Consult Clemson poultry scientist
B. Consult patient's transplant surgeon
Outcome: Risk of multiple poultry viruses and microbes is too great for the immuno-suppressed patient

II

Client/Patient: Realtor bitten on index finger by spider on sunshield in her car
Question: Why is the finger swelled and inflamed despite antibiotics?
Methodology:
A. Receive color-photo of lesion promptly
B. Prescribe Dapsone, an anti-inflammatory, antinarcotics medication, orally for brown recluse spider venom, to supplement the antibiotic.
Outcome: Rapid healing with little scarring

III

Client/Patient: Orthopedic surgeon recovering from severe giardial infection (diarrhea)
Question: How and why did he acquire a giardial parasite? Exposed to cats while fishing and dogs while hunting
Methodology:
A. Obtain an accurate history of the onset of symptoms and determine known incubation period
B. Establish that dates coincide with office visits of two infants for plaster cast-removal (pelvic area/soiled with feces)
C. Interview both mothers to confirm the infants were still symptomatic with giardiasis at the time of their office visits
Outcome: Source of giardia exposure is most likely inhaling plaster dust while sawing off fecal-contaminated pelvic casts from two symptomatic infant giardia-carriers. Dogs and cats exonerated. Case documented to CDC (first reported case of pediatric aerosol transmission to MD, 1981)

(headaches, insomnia, sinusitis, unusual fatigue, and paresthesias). Despite a lack of signs of organic illness, Dr. Brown felt uneasy about the timing and onset of her multiple symptoms. He was treating her for long-standing anxiety and depression with counseling, but he felt that her story warranted liver-function tests and another blood count for possible anemia. Both were reassuringly normal. At the point of presenting lab results to Ms. Evans, she was furious and challenged him to do further tests, as she was searching the Web for clues to her "environmental" illness. Somewhat nonplussed at this very intelligent, articulate, demanding patient, Dr. Brown called the chemist at our agromedical pesticide-residue lab for the possibility of running her serum for traces of organochlorines or other pesticide residue. In fact, the clinician felt that her data base for environmental illness exceeded his own primary-care training.

When the serum tests were completely negative (even for parts per billion of residue above standard population norms), Dr. Brown was as surprised as Ms. Evans. Not unexpectedly, she grew angry again, challenged his diagnostic skills, and announced that she would switch doctors.

Follow-Up

Soon after, Dr. Brown presented her case at grand rounds for input from his teaching colleagues; they found no fault in his patient, caring, and persistent approach to this complicated case. In fact, they were glad that they had few encounters with such formidably aggressive patients who are enduring somaticizing symptoms with an overlay of chemophobia. Over time, the brilliant but unfortunate journalist, Ms. Evans, developed a series of severe organic illnesses requiring several major operations and chemotherapy. She went through a series of clinicians who had to cope with her angry personality, high IQ, and recurring somaticizing, multisystem complaints. Dr. Brown recovered from his disappointing encounter, licking his health-provider wounds, and was motivated to write up his experience for the *Journal of Agromedicine,* to share with other MDs who may encounter similar difficult cases. In fact, Dr. Brown served with the author on a panel of South Carolina teachers of family medicine to develop guidelines for taking a simplified, practical occupational/environmental medical history from adult patients entering a family practice clinic.

Taking a WHACS History in a Busy Clinical Practice

WHACS is a helpful acronym for the primary care physician. It reminds the physician to ask five leading questions on occupational/environmental exposures; with practice, this interview should take five to ten minutes.

W - What do you do as an occupation?
H - How do you do it (a typical day)?
A - Any health concerns (about your work)?
C - Any co-workers' health concerns?
S - Are you satisfied or dissatisfied with your work (level of stress)?

Where would the medical profession be, entrenched in old habits, without the challenge of brilliant adversarial patients to test their mettle, their resilience, their empathy, or their ongoing education in clinical experience? Thank you, Ms. Evans!

Figure 2.2 illustrates the simplified choices for differential diagnosis of occupational and environmental medicine (OEM)-related illness. The task is by no means as simple as it appears. However, the four choices dictate the management in each case. It takes an astute clinician to handle the complex history, physical exam, and lab tests of a presumably exposed and damaged patient in the average busy ambulatory setting of a primary care practice. But with time, the clinician will gain confidence in handling or referring cases of OEM-related illness.

Illness present	Exposure present	Diagnosis
Yes	Yes	Positive
No	Yes	Monitor the patient
Yes	No	Negative
No	No	Reassure the patient

FIGURE 2.2. Differential Diagnosis and Management of OEM Disorders

Summary: No Single Path to Discovery

The foregoing examples of agromedical cases are cited here to prepare the reader for the common denominators, the basic methodology, and the essential tools required to investigate a range of real-world problems encountered by the author in over twenty years of clinical-epidemiological experience.

Little in my specialty training in pediatrics prepared me for most of the complex adult illnesses. Little in my public health training in epidemiology prepared me for the pharmaco-toxicology of pesticide poisoning. Little of my experience in private practice prepared me for challenges in medical entomology. Little of my academic duties prepared me for statewide consultations with extension agents, clients of extension, and agrichemical industry professionals during a committed forty-hour week of statewide public service. Yet the truth is that I needed every lesson learned from a variety of educational experiences to be useful as an agromedicine physician.

Somehow, patients know better than young doctors that their healers are on a learning curve with their fair share of hits and misses, cures and non-cures. That's why they call it "medical practice": practice, practice until it comes out right.

So it may be with the learning curve in agromedical investigation. It requires practice, practice until one begins to use a variety of tools, strategies, and tactics in agromedicine. With a little luck and serendipity, one may find a rainbow in an ordinary washtub.

This brief introduction to the strategy and tactical tools of agromedicine should enable the reader to follow the threads of interdisciplinary reasoning, hypothesis-testing, and differential diagnosis involved in the following narratives of client-based problem solving.

RESOURCES

The following references that I have authored or co-authored may help the reader delve a bit deeper into the art and science of finding rainbows in washtubs.

Agromedicine: Selected Papers from the First Ten Years of the Journal of Agromedicine. 2005, The Haworth Medical Press.
Environmental Epidemiology for the Busy Clinician. 1997, Harwood Academic Publishers.
Practice-Based Epidemiology: An Introduction. 1986, Gordon-Breach Publishers.
User's Guide to Agromedicine. 2000, Harwood Academic Publishers.

Chapter 3

Nikes by the Nile: Agromedicine in Alexandria

SCENE ONE

"Hey, Stan, guess what? Your outline for the Alexandria project looks okay. It was approved by the research office. Is your passport up to date?"

That is how I recall the collegial conversation with my dean, who needed, and found, a faculty member to use the AID (Agency for International Development) grant-supported travel the university had received.

The next thing I knew, I was wending my way through the Cairo Airport, surrounded by tourists, armed security, porters, taxi drivers, immigration officials, and an abundance of well-fed cats (used to control rodents!).

Then I was off to Alexandria by car, driven by a university-designated driver. I checked into a modest hotel near the campus of the University of Alexandria. The hotel is actually one of former King Farouk's summer palaces on the Mediterranean, updated around 1950, used by the medical university to house visiting faculty.

My ag-med proposal, written with a minimum of available research data, assumed that irrigated fields and farms (near Nasser's desert reclamation project) would be a worthwhile site for agro-

Rainbows in Washtubs

doi:10.1300/5851_03

medical investigation. Surely there would be health concerns among the row-crop workers who labored in fields of cotton, flax, wheat, sorghum, and, perhaps, fruits and vegetables. After all, this delta region was once the bread-basket of the Roman Empire. Now the delta consisted of thousands of acres of abandoned desert. Hopefully, it could be reclaimed by an elaborate Nile-based irrigation system.

The afternoon Mediterranean sun beckoned me to wander up a few promising streets along the historic Alexandrian shoreline. A rocky, litter-strewn beach bordered a heavily trafficked highway. Honking trucks and lorries filled the air with diesel fumes. It was hard for me to imagine Alexander the Great falling in love with this harbor and coastline and choosing this site for his capital city.

The next morning, over a breakfast of ultra-strong tea and goat cheese–laden toast, I wondered what I could do to be useful to delta farmers. How could I foster a joint university effort in agromedicine in Egypt?

Then I recalled that several prosperous tomato farmers on Hope Island back in South Carolina had invested in a small vegetable farm outside Alexandria. It was designed and managed by several California farm technicians who were experts in desert agriculture. Luckily, I had saved a phone number in my briefcase, and after several failed attempts to place a call on the single antique phone in the ancient marble hotel foyer, I got through to Dave. His friendly southwestern voice welcomed me to visit the demonstration farm that he was so proud of. Would I be ready to be picked up in an hour or so? He would be coming in a dusty, red Chevy Blazer. I grabbed my sunglasses, sunhat, camera, and work shoes, looking forward to finally seeing one of the farms that I had come so far to observe.

From under the brim of his sweat-ringed Stetson, I could see Dave's blue eyes and weather-worn smile. We bustled though some high-speed traffic with macho drivers and continuously honking horns. The sun-bleached delta desert stretched for miles beyond the perimeter of Alexandria, until we came to lush green irrigated fields here and there. Scattered about were a few humble dwellings with plenty of bird roosts. Dave explained that pigeons and gulls make good eating for some folks.

At last, an hour later, we pulled up into a beautiful, symmetrically arranged field of irrigated vegetables. The rows of ripening tomatoes, budding asparagus, beautiful avocadoes, and vibrant lettuce would impress anyone. Dave pointed out to me that under the desert sand layer, the roots can tap some of the richest topsoil you can imagine. Shiny, new imported John Deere tractors were plowing fresh acreage in adjoining fields. (In addition to monetary investments, the Hope Island entrepreneurs also had a contract to sell and service tractors and other farm equipment to the Egyptian farmers.) All of the fields were protected by barbed wire to keep out intruders and night-time-poachers.

Dave's office was a simple cement-block, air-conditioned unit with its own computers and weather monitoring equipment. After some cold refreshments, I was guided to a huge hangar-like packing and shipping facility, air-conditioned and bustling with belts carrying crates of fresh vegetables to a large waiting air express truck. These were headed to waiting markets in France, Switzerland, and Italy, Dave explained. Air-freight costs here were minimal, because the empty space on planes is simply waiting for Egyptian exports, such as these!

"The key to the markets is *quality,*" Dave explained, "which we take seriously. In fact, our tomatoes brought home blue ribbons in competition last spring in Brussels!" My tour with my host led me past rows and rows of young Egyptian women busily and dexterously packing the cartons for air shipment on a conveyor belt to the waiting refrigerator trailer-truck, destined for the Alexandria airport. In their native costumes of knitted sweaters and colorful bandanas, they wore warm gloves as protection against the interior's air-conditioning. They were also wide-eyed at my visit with Dave, laughing to each other about the stranger with his camera and notebook.

Next we had a snack for lunch in the office where I was introduced to Omar, Dave's Egyptian assistant. He smiled modestly as Dave heaped praise on his energy and enthusiasm for this work. Now it was time for Omar to lead me into the fields where the busy hands were planting and staking the vines in the sunny dry desert heat. Here there were rows upon rows of very young workers, actually girls in their teens. They wore loose flowing layers of cotton and colorful bandanas, which blew gently in the occasional desert breeze. They also worked cheerfully and laughed at me, the visitor with the camera.

Each row of rich, dark, tilled delta soil was moistened with rows of plastic tubing feeding essential moisture, nutrients, herbicides, insecticides, and fungicides to each plant, delivered by the massive irrigation network, which was programmed by a computer. *This was truly high-tech,* I thought, *worthy of the best U.S. fields I had seen.* Omar told me that he hoped to take Dave up on a promised visit to his friends' farms in Arizona to learn more about plant selection, crop protection, and mechanical innovations.

On the ride back to my hotel, Dave confided that his investor-entrepreneurs hailed from Connecticut and California, as well as Hope Island, SC. They were attracted by the low tax rates, incentives to farm by Cairo government, cheap labor, and inexpensive shipping for the products. Not to mention the *first* asparagus operation in the delta for over thirty years! The next day, and the next, I tried to observe *something,* anything worthy of a well-designed epidemiologic study.

My 35 mm camera captured some of the culture shock I felt and wanted to carry home to share with medical students at Charleston: the beautiful light, the endless horizon, the lush plants, and the rows of female children laboring in the fields while their mothers and older sisters worked in the packing house. Then my eye caught the tall figure of an elderly man in a black robe and turban, carrying a long willow switch as he monitored each row of child laborers. "That is the culture," Omar explained, and I thought to myself, *not unlike the abusive, crew chiefs supervising many migrant worker operations in the United States.*

What am I doing here? I asked myself. This is not what I envisioned in Charleston at the medical center as a clear-cut, well-designed, laboratory-assisted field worker study coordinating two motivated universities (Alexandria and Charleston).

After several trips to and from the farm with Dave, the dusty, desert routing became monotonous. Trying to avoid the inevitable growing awareness of how little I had to offer on this pilot, modern-appearing, California-like operation, I looked again at the children laboring between the damp earth-soaked rows of vegetables—for an intense moment I *really* looked. And there it was; a true agromedicine problem, that was easily diagnosed, measurable, and preventable.

The girls were working in the fields *barefoot!* Their colorful, open-toed sandals were carefully stacked at the gate to the fields, to be preserved until the end of their workday. Aha! What about skin absorption of any or all of the variety of chemical treatments to the irrigated soil? Toxicologically, this hourly exposure could be as toxic as aerial spray or ground-rig spray to unprotected workers. Skin absorption would be subtle, but cumulative, especially through any abrasions in the vascular skin of the foot or ankle. With a little imagination, I realized that I had something to offer in this strange (exotic for me) setting—protective footwear!

SCENE TWO

Ultimately, my proposal of a real, short-term pilot study to improve worker health would not become a reality. While the farm manager and the farm workers were cooperative with my research, I encountered many difficulties when working with the local university faculty. In the end, the personalities, the politics, and the conflicting priorities of all involved created obstacles that could not be overcome. In addition to my goal of fostering a joint university effort in agromedicine, some wanted grain dollars, originally AID funds, for pesticide research and student fellowships. Others were interested in funding international faculty exchanges. In the end, teamwork and compromise between the players were not possible.

SUMMARY

What had I learned (at age 55) from my trip to Egypt? Here are a few lasting impressions:

1. *The latest agricultural technologies* (such as desert-irrigated crops developed in Israel and California) are highly feasible and exportable to settings like the delta, providing food, jobs, income, and quality of life for needy workers.
2. *Agromedical scientists* really enjoy working together. I worked with an agriculture professor who could not have been more helpful, sincere, and hospitable. However, the social-prestige gap between the medical school and the agriculture school within the same university was as great as in some of the state-funded schools in the United States.
3. *A site visit* with repeated observations can ultimately save the investigator from wasting time and energy on a fruitless endeavor, a theoretical plan for a research project that won't work in a different country or in a different culture.
4. *A fresh perspective* on the importance of the dictum: "Think globally, act locally." It was clear to me that back home in South Carolina, two universities would have to work together, the agriculture college and the medical college, in a way that the Egyptian faculties resisted. Unless we learned locally, in South Carolina, to blend agriculture and medical skills, we were not qualified, or ready, to think we could execute or export agromedicine internationally.

5. *My memories* of dark-skinned girls smiling for my camera, with wonderfully colored bandanas, flashing dark eyes, giggling and laughing at the stranger from a foreign land are still with me. Child labor in agriculture is prevalent and difficult to overcome.
6. *My imagination* carries me back to the civilization of the Nile, built on the backs of peasant workers tilling the rich alluvial soil, growing an abundance of crops and food to feed an army, an empire, a pharaoh's entourage of priests, and plenty for the Voyage of the Dead in the tombs that became Wonders of the World. Who can forget how Nile-nurtured grain that Joseph stored for seven fateful years managed to ward off starvation during the next seven years, to save the Pharaoh from disaster?

FOOTNOTE

I forgot to mention my near-disaster at a pool, where I went to swim a few laps. Emerging after a half-hour swim, I was almost run over by over-sized Mediterranean harbor rats, scurrying and scavenging around the edges of the pool.

Back in the United States, my epidemiologist-mind fretted over my calendar for the entire three-week incubation period for leptospirosis, which I was sure I had caught from rodent urine. This would be some kind of "souvenir" of my Alexandria adventure.

Chapter 4

Dangerous Beauty: Peruvian Lily

SCENE ONE

"Merry Christmas, Stan! Merry Christmas, Joanna!" Judy's warm greeting at the door carried all the holiday cheer of the season. Little did I know that within the next ten minutes Judy's handshake would start me on another agromedicine adventure.

As Judy offered me fruitcake and punch, I couldn't help but notice her right hand. The thumb and index finger were swollen, red, peeling—almost painful to look at. Her left hand was normal. *How come?* I thought to myself. As the joyful reunion evening waned, with most of the other guests departing, I couldn't help speaking to Judy who was cleaning punch cups in the kitchen.

"That's quite a rash you have there, Judy. You really need to get it treated. It looks like a combination of allergy, irritation, and secondary infection, possibly fungal."

"Doc," she said, "tell me about it! I have dermatology bills and pharmacy bills from here to Gazoo! Dr. Davis is about to give up on me, trying every new form of cortisone. She thinks that stress may be contributing; you know, they call it neurodermatitis? What do you think?"

Rainbows in Washtubs

doi:10.1300/5851_04

"Honestly, Judy, I wouldn't know. Dr. Davis is the best dermatologist in the Low Country—university trained!"

SCENE TWO

After the holidays, medical students' classes resumed. One day, we escorted a dozen juniors to Clemson University's Vegetable Research Lab. I hoped they would be more interested than the last group. Dr. Jason was really excited, showing them how to tissue-culture new cultivars of tomatoes (heat and humidity resistant) in Petri dishes. This new technique accelerated the breeding of prize tomatoes. It leapfrogged over traditional slow methods of genetic engineering.

In a short time, in Dr. Jason's lab, the front row of students showed some interest in the methodology. The students in the back row were bored as only medical students can be bored; they couldn't wait for lunch. A tractor-drawn ride around the rows of experimental vegetable plots helped, thanks to Dr. Jason's over-friendly collie cheering them up. Meanwhile, I chose to kill time in the air-conditioned lab, flipping through horticulture journals, until the students' finished their field tour.

Several issues of a floral industry trade journal caught my eye, with their beautiful colored photos of new varieties of floral decorative and landscape plants. Roses were prominent, along with climbing ivy, decorative cabbages, and varieties of lily, domestic and imported species. One photo caught my eye because it resembled an orchid to my untrained eye. It was called alstroemeria or Peruvian lily. What a spectrum of colored varieties, one more beautiful than the next!

A footnote stated that an NIH scientist (dermatologist) warns readers that some imported flowers can cause *severe* dermatitis, resembling the well-known tulip bulb rash, a chronic inflammation of the thumb and forefinger of certain susceptible tulip-bulb packers, discovered years ago in the Netherlands.

SCENE THREE

Back at my toxicology lab, my administrative assistant, Tom, asked how I enjoyed the office party at his home. I assured him that Joanna and I enjoyed the evening immensely. Joanna especially raved about the gorgeous floral arrangements on the mantles and the refreshment table, and the sun porch which overflowed with tropical plants.

"Oh, didn't you know Judy is an expert floral designer? She is in demand for weddings, showers, and funerals, especially when the commercial florists are overwhelmed. Judy can work when she feels like it, in the sunroom with her own little work area. The extra income doesn't hurt our budget, either."

"Does her rash slow her down, Tom?"

"Oh yes, but she loves her work."

Thinking of Judy's rash and the floral magazine photos and footnote, I decided to do some research in the library on botanicals causing contact dermatitis. This was definitely *not* my area of medical expertise. I finally decided to contact Dr. Erwin at Bethesda to elaborate on his cautionary note in the magazine.

SCENE FOUR

"Dr. Erwin, this is Stan Schuman calling from MUSC in Charleston. Can you help me figure out a severe form of contact dermatitis troubling the wife of my office manager? She has baffled one of our best dermatologists here, after months of treatment. I'm wondering from your footnote whether her part-time work as a floral designer could explain her rash, and why it resists treatment. Judy doesn't handle, pack, or sort tulip bulbs."

"Stan," he said, "you should ask Judy about handling alstroemeria. I have several patients that I'm writing up for the academy journal. I can send you several pre-publication photos of a typical rash on the thumb and forefinger. Also their skin-patch test results."

SCENE FIVE

As my interest grew from this first case, I visited several busy floral shops in the neighborhood. The friendly owners seemed pleased that an MUSC physician was showing interest in their craft and their design skills. "I am impressed," I said, "by how much goes into designing and arranging floral bouquets, centerpieces, corsages, and such."

"Yes," the proprietor said. "It takes skill and patience. But now, several of our best designers are complaining about their medical bills. This new kind of rash won't heal. Even wearing cotton gloves doesn't help. One of our best employees just had to give up her work here completely."

I could hardly wait to ask, "Do you work with alstroemeria here?"

"Oh yes, doctor, they're the latest, the most popular lily in years! Imported from Holland at first and now from Latin America. Look, doctor, did you ever see such a variety of colors? They're the poor man's orchid. And they're so affordable. Our customers love alstro, and we can get it for them all year-round."

SCENE SIX

I made a house call to Judy (with her MD's approval). Forty-eight hours later, Tom followed the methodology of simple skin patch testing, outlined for us by Dr. Erwin at Bethesda in his preprint article:

1. Cut-out pieces of flower (leaf, petal, and stem) are applied under special allergen-free Band-Aids.
2. A Vaseline "control" is applied under another Band-Aid.
3. Several suspect flowers or floral chemicals can be tested on cleansed skin between shoulder blades.
4. Showering must be avoided for forty-eight hours; any itching areas must not be scratched or irritated.
5. Delayed hypersensitivity will be revealed, when patch tests are removed.

Holding my breath, I waited the required forty-eight hours to revisit Judy. Before I could enter her front door, she complained, "My back is itching like crazy! I've *never* been allergic to Band-Aids like this before!"

"I'm sorry, Judy, but we had to wait forty-eight hours."

"Stan, are you sure that this research can actually help floral designers like me?"

It didn't take an expert trained in dermatology to read Judy's patch tests. A sophomore medical student could do it.

Here are the patch-test results on Judy's back:

Floral freshener solution	Negative
Floral putty for stems	Negative
Baby's breath	Negative
Decorative fern	Negative
Rose petal	Negative
Alstro leaf	Positive
Alstro stem	Positive
Alstro flower	Positive

The redness and swelling of each positive reaction was palpable to the touch just like a four-plus tuberculin (delayed hypersensitivity reaction). The rest was simple. No wonder any re-exposure triggered her immune response. These tests proved that Judy and her fellow florists did not have neurodermatitis. Avoidance of alstro, once diagnosed, was the only "cure." Even rubber gloves do not work. The alstro chemicals could permeate rubber (but not synthetic nitrile gloves, according to Dr. Erwin in Bethesda).

The finishing touch: A statewide survey of South Carolina floral designers revealed an upsurge in severe contact dermatitis (affecting over 12 percent of respondents). Careful histories revealed that the onset of the severe rashes coincided with the rising popularity of alstroemeria among florists, designers, and the public.

Prevention

- How can exposure be avoided? Every knife, pair of scissors, work surface, and vase carries molecular doses of alstroemeria sap (alpha butyrolactone). It is as powerful as any plant sensitizer (poison ivy included).
- Once sensitized, there is no cure. A person must avoid even casual contact with the flowers.
- How can the problem be abated by floral managers? Isolate alstro to one small work area for use by nonsensitive designers.
- More managers simply stop importing or handling the offending plant in their shops.

Interesting Sidelights

- Some florists have told me that exporters from Latin America stash cocaine and marijuana in boxes of plants bound for the United States.
- Harvesters seem to handle the crop without side effects (perhaps because it is machine cut and bundled for export). The highest risk for contact dermatitis occurs among floral designers who trim stems, using the thumb as a backstop, and

who thread wires in the stems making their flower arrangements.

- The Dutch horticulture industry exported the exciting new variety to the United States floral industry without due warnings of the side effects (tulip-like finger dermatitis), which their own dermatologists and chemists had documented in the 1930s (Slob et al., 1975).
- Question: Why didn't the keen diagnosticians in dermatology practice make the connection between alstro and resistant dermatitis? Answer: Probably the rarity of floral designers (small retail operations) in a large private practice made neurodermatitis a more likely diagnostic option. In truth, the bills and frustration caused by the easily infected rash would make anyone, including Judy, feel desperate.

REFERENCE

Slob A, Jekel B, de Jong B, Schlatmann E. On the Occurrence of Tuliposides in the Liliflorae. *Phytochemistry* 1975;14:1997-2005.

Chapter 5

Dyrene Mystery: Migrant Workers at Risk

SCENE ONE

"Stan," Professor French said, "it looks like you've done a good field study in Tennessee. Your data shows attack rates, time of onset of rash, and relation to lack of protective clothing. Your chemist shows negative results for overdose of chemicals on the tomato leaves and the worker's blood sera. Your time chart of events shows good history-taking, *but* guess what?" A long pause ensued, and I held my breath. "Your clinical diagnosis is all wrong. Where in the world did you study dermatology? This cannot be a chemical irritant dermatitis."

"Why not, Dick?" I asked in a state of shock.

"Because the clinical picture simply doesn't fit. The rash spared the face, and took forty-eight hours to erupt. In fact, with a second visit to the farm, you can prove the correct diagnosis—delayed allergic contact dermatitis—with my help, of course."

"You can't be serious, Dick. That small town was so hostile to me, to the health department, to any outsider. I don't know what they might do to me next time. I feel sorry for the migrant workers.

doi:10.1300/5851_05

They must be anxious to finish up and clear out of those farms, as fast as they can."

"Well, do you want the truth or not?" French insisted.

"Okay, okay, tell me again what's wrong with my work-up?"

The next thing I know, I'm following French's detailed instructions. I am working in the lab with Gary, our chemist, mixing 1:100,000 dilutions of three of the eleven agrichemicals used heavily on the tomato farm (benomyl, endothal, and Dyrene). We are following the guidelines of Professor French, as if we were preparing antigens for skin testing in his dermatology clinic. Together, Gary and I prepared enough allergy-free adhesive strips to skin test a dozen of the affected migrant workers. Then, against my deepest doubts, but respecting the acumen of Dr. French, I had to pack up to return to east Tennessee. I needed to finish the investigation. I had to carry out a delicate diagnostic contact-dermatitis, patch-test routine usually done by trained nurses in a pristine medical office, not in a field filled with crops.

SCENE TWO

So how did my story of dermatologic epidemiology develop in the first place? Actually, it started as a fairly routine phone call for assistance from a young PHS physician serving his payback time at a remote, rural clinic in the foothills of Appalachia, Tennessee. The EPA offices in Tennessee, both regional and state, had punted, forwarding the call to the Washington, DC, Office of Pesticide Regulation. That office dumped the call to our medical center, an EPA supported laboratory. This was fairly standard practice for the DC office to refer difficult calls to regional EPA grantees.

The angry frustration of young Dr. Zed was evident from the first hello. "What are you fellows at EPA getting paid for? This is an outrage! Last weekend I had to treat not one or two, but a *dozen* poor migrant farm workers. They had a terrible rash, a chemical burn actually, on the arms of both the men and women, and on the chests of the men as well. Luckily, I had enough cortisone and antibiotics to give them each a shot to get them back to normal, I hope. I'm positive that they were deliberately sprayed by a ground rig, run by the farmers and his foreman, while they were busy picking and staking the plants. What can be done to punish the farmers? I've called every office in the county and state, but they won't touch it. The DC office referred me to you. You will probably give me the same runaround, too. My next call will be to the Environmental Defense Fund. Their lawyers know how to take action! Besides, I have color photos of the burned skin areas for the record."

"Wait a minute, Dr. Zed. Cool down," I said. "Give me the history again, a little more slowly." Minutes later, it was clear that this was an outbreak of some type of agriculture-related severe dermatitis, requiring more than usual topical creams and ointments. How could there be eleven to twelve workers all erupting at the same time, all during the same weekend?

The adversarial/activist aspects of the outbreak did not interest me beyond ordinary sympathy for any worker unfortunate enough to be afflicted. But, it struck me as a classic "true" epidermiologic series of questions: Why these patients, why this common exposure, why this time of onset, and why such a similar clinical presentation (severe rash of exposed, unprotected forearms and bare chest during hot September weather)? I urged Dr. Zed to send me, post-haste, copies of color photos he took in the clinic on late Saturday and Sunday nights in Tennessee.

When the photos arrived, there was no doubt that the rash was as severe (peeling, inflammation, blisters, and excoriations) as Dr. Zed promised. On the phone again, Dr. Zed shrewdly inquired, "Do you have a chemical lab to test their blood for pesticide poisoning?" I assured him that our lab was one of EPA's premier regional reference labs for testing pesticide residue. "Good!" he replied. "I've got to get those bastards! And help these poor migrants. They work so hard and don't complain. If they do they get shipped back to Mexico."

SCENE THREE

So, unable to say no to Dr. Zed, here I was in the foothills of Tennessee. My mission is to try to collect human sera for testing. In addition, I had one inexperienced assistant carrying four dozen chemically clean jars in a government station wagon, to help me collect punches of tomato leaves from the suspect fields. Some of the plant samples could provide the evidence of pesticide misuse by the farmer, based on chromatographic analysis at our lab in Charleston.

Near sundown, I arrived at one of the cinder block huts of the field hands. I tried to convince the matriarch of the extended family of Mexican workers, that:

1. I am really a physician,
2. I am here to help,
3. I need blood samples from each of her sons, daughters, and daughters-in-law, and
4. despite the EPA logo on the station wagon, I am *not* interested in checking their immigration status.

Worst of all, I could neither speak, nor understand Spanish. Luckily, Mrs. Gomez had enough English to help me. The family had been painting houses in Houston for several years. This was their first row-crop job, and it was not turning out well. The crew chief was drunk all the time; he would probably cheat them on pay day and pocket part of their wages or threaten to turn them in as illegals.

September dusk was setting in and it was time to leave. I paused at the door knowing that tomorrow I would try to persuade Mrs.

Gomez (and the extended family) to cooperate in the investigation. Several grandchildren and her own baby kept Mrs. Gomez busy, as she somehow managed to prepare supper. The field hands would be returning soon from the farm. Not knowing what, if anything, tomorrow would bring, I tried to let go of it all—yesterday, today, *and* tomorrow.

I began to inhale the aroma of chili pepper, seasoned tortillas, and fried beans and pork cooked on the primitive stove. In no hurry, I began to look around at the neatness of well-made bunks, the swept concrete slab floor, a prominent crucifix, folded laundry, and shining kitchen utensils. "Mrs. Gomez you're doing a great job here. These are really tough conditions," I said to her sincerely.

I searched her face to seek her expression. She met my gaze for the first time. After a pause, in nigh-perfect English, she asked, "Would you like one of my tortillas?"

I said, "Yes ma'am, they sure smell good!" And that was how we two strangers crossed the cultural divide; my hunger and her cooking saved the day.

SCENE FOUR

The next morning's collection of blood samples (for chemical residues) from the migrant workers went unexpectedly smoothly. No complaints. Even my primitive non-Spanish history-taking (diagrams and hand signals) yielded reasonable spoken information from each of the seventeen workers. I needed information on the timing of rashes, exposure to field ground rig treatments, protective clothing, which crop was being worked, and so forth. Even the punch-leaf collections by the assistant went smoothly. That evening we headed back to Charleston with all the blood and leaf specimens and histories we came to collect.

We felt sure that we could use the lab analysis to document unintentional chemical overexposure and consequent chemical irritant dermatitis, as excitedly reported by the young Dr. Zed. Until Dr. French, our distinguished professor of dermatology, upset our neat theory completely. He forced us to reconsider the diagnosis accepted by the angry young doctor, and accepted by us, because we knew that pesticide misuse was not an infrequent occurrence: inappropriate, accidental, or careless overexposure of field hands.

SCENE FIVE

Over three weeks had passed. Autumn colors glorified the Tennessee foothills, as I drove back alone this time. I hoped to find the migrant workers before the last crop was harvested. They would be picking tomatoes now in adjoining fields. Not much had changed. The rashes had healed with some peeling. Half of the migrant field hands were gone. But the extended family of Mrs. Gomez remained.

Fortunately, the weather had cooled so that long sleeves were comfortable for the workers (less skin contact). I managed to find seven remaining workers to apply the patch tests (endothal, benomyl, Dyrene, and Vaseline control) after careful cleansing of the scapular area of skin. I explained Dr. French's routine, no bathing for two days, I would return after forty-eight hours to read the tests. Mrs. Gomez and one daughter-in-law (neither exposed to field work) served as controls. The next two days it rained. Alone in my austere motel room, I spent the time charting the histories, outlining the timing of possible exposures, and tabulating the negative lab results from plants and sera for excess chemical residue.

Finally, early on the third day, the daughter-in-law picked me up in a battered pickup truck to drive me to the distant tomato fields where the family was now working. Even in the rain, the beauty of Tennessee caught my breath.

We actually passed two small waterfalls, on several privately owned properties. I took Kodachromes with my trusty 35 mm camera, which had documented some of Dr. Zed's findings on my first trip. Now, I needed to photograph each of the patch test results for Dr. French's demanding protocol. We found the tomato fields and the workers. They took a break to come to the truck. We lifted shifts and blouses to reveal the patch tests. I hoped the morning clouds would permit enough color detail to satisfy Dr. French's critical eye. (He said he didn't trust me or any untrained RN to do the testing right!)

One by one, the tests revealed exactly what they should: delayed hypersensitivity, raised, reddened lumps around the 1:100,000 diluted chemical, and no reaction to the Vaseline control. I couldn't believe what I was seeing. How unlikely for our field skin tests to single out one of eleven possible chemicals used on the farm. How unlikely that the chief suspect chemical, endothal, would be negative on every skin test. Instead the long shot, the fungicide Dyrene, used only intermittently depending on mold infestations, proved indisputably positive in six of seven persons available for testing.

Dr. French was right. The rash that spared the face and waited forty-eight hours to erupt was *not* a direct chemical irritant type of dermatitis. Only delayed hypersensitivity to an allergenic compound like Dyrene (or poison ivy or poison oak antigens) could cause this pattern and this type of prolonged rash in an exposed group of workers.

The accused farmer-owner of the fields, and his foreman son, were exonerated of causing an overspray (premature entry to treated fields violation). They were innocent, at least medically, of causing this type of outbreak of rashes. The medical reason, as predicted by Dr. French, is that even tiny doses of an antigenic chemical can cause hypersensitivity to erupt at least forty-eight hours later, as a florid, severely itching contact dermatitis.

SCENE SIX

The solitary, long drive home in the rain from Tennessee to the Low Country of South Carolina gave me time to ponder my gleanings from this investigation:

- I had learned some basic clinical dermatology—better late than never—from a master dermatologist, Dr. French.
- I had earned the trust of one migrant family and its matriarch, even if they didn't understand what it was all about.
- I helped the farmer, his son, and the local doctor, who had encountered sporadic outbreaks in prior tomato season. He now had a diagnosis to work with.
- I had responded to a sharp young MD's basic epidemiological question: Why did all these patients appear on one weekend, from a single possible environmental exposure?
- I would have a timely report for an interested journal (perhaps *Lancet*) to document a previously undocumented allergic risk to a widely used fungicide. This would prevent the kind of missed diagnosis by young Dr. Zed and me, as well as the conscientious local doctor who was baffled over several summers of treating migrant workers' dermatitis.

I had a good laugh, driving home, recalling the final farewell scene at the only breakfast café in town that morning. Here I was, seated by chance, right next to the farmer, his son, and, believe it or not, the regional agrichemical Dyrene salesman. Finishing their grits, gravy, and country sausage, they had to lean across their table to tell me I had been wasting my time: "Why waste time to study common tomato rashes?"

All of a sudden, the father and son raised their sweat-stained T-shirts to show me the itching, peeling, healing stage of dermatitis around the waistline that they called "ordinary tomato vine rash."

In fact, I observed, their rash was identical to that of their field workers.

SCENE SEVEN

Two years later, I am seated in the medical school auditorium of the University of Alexandria, Egypt waiting my turn to present slides on agromedicine teamwork between Clemson and MUSC. Seated next to me are WHO health workers, agricultural science faculty from Cairo and the health department. Speakers on toxicology of pesticides, integrated pest management, and biotechnologic research are introduced.

At the refreshment break, a pleasant looking, scholarly attendee studies my name tag and introduces himself as an experimental pathologist for Mobay Inc., the maker of Dyrene.

"Are you the same Dr. Schuman who blasted Dyrene in a *Lancet* article?"

"Yes, I am, sir, along with Dr. French."

"Well, you're both wrong, you know. And it seems that our registration with the EPA will not be renewed. Growers really depended on using it. There aren't too many effective fungicides available."

"Why were we wrong?" I asked.

"Because our standard eye and skin tests on several strains of rabbits were consistently *negative* for any Dyrene irritation."

"I'm sorry our work deprived you of a good product, Doctor, but hyperimmune contact sensitivity can only be prevented by strict avoidance, especially if a person is once sensitized. In fact, you know, our lymphocytes carry hypersensitivity in memory for the rest of our lives. It's the same kind of antigenic-mediated response as tuberculin or poison ivy."

"Humph," he growled at me as he stalked away. I realized that this expert toxicologist was still unconvinced and unimpressed with my explanation.

He would not, could not, acknowledge a difference between humans and rabbits in their immune response, evident in this unusual case of Dyrene.

Chapter 6

Parasitosis: The Enemy Within

There are many gaps in medical education, medical theory, and medical practice. Unfortunately, the average patient would rather not think about the inevitable truth. One has to endure the unintentional mistakes in diagnosis, treatment, and prevention that occur from time to time. Even the best major league batters do not expect to bat over .300 every season, and there are few twenty-game-winning pitchers.

Take, for example, medical entomology. Consider the vast world of insects that inhabit this planet with us. From early humans with unrelieved lice to the modern day fight against mosquitoborne diseases (malaria, yellow fever, West Nile virus, filariasis, Dengue, and a whole range of hemorrhagic fevers), we tend to forget how little we know about early detection and containment of insect-related maladies. European civilization was brought to a standstill because of the Black Death, carried and spread by the rat flea for centuries.

Here are two small stories from our agromedicine file: the first, a case of nonparasitosis and the second, a case of true parasitosis. In each case, the patient suffered as much from the inept medical profession as from the parasite problem itself.

Rainbows in Washtubs

doi:10.1300/5851_06

SCENE ONE

Nonparasitosis

It was not unusual for our office to receive calls for help or advice from Clemson extension agents around our agriculturally diverse state. This call came from the agent of a mid-state county rich in cotton and tobacco. "Doctor, I'm really concerned about the health of my pastor and his wife and their children. Their home seems to be infested with some kind of bug. They are not engaged in farming."

Such is the diverse influx of calls for help which a broadly trained, experienced county extension agent is likely to receive, especially while serving a rural, close-knit community. On-call in their profession, the county agent adds a whole dimension to the term *public service.* The clients of Clemson Extension expect answers to a great variety of agricultural, food safety, gardening, and health concerns.

The pastor's wife, Donna, a popular and active member of the church and community leader, called Hal early Monday morning to bring in some special insect specimens. "These parasites have been crawling under my scalp and my skin for weeks. Now, they're spreading to my husband and our teenage sons. These creatures are incredibly resistant to anything we do to eliminate them."

She then described weeks of intense house cleaning, carpet cleaning, sanitizing and scrubbing, and boiling of the bedding, not to omit repeated chemical treatments by expensive pest control professionals. Her home was less than three years old, but Donna and her family, who had lived up north in Vermont, had never experienced the cockroaches, fire ants, or ticks bothering them in South Carolina.

Donna knew that Clemson University has a highly regarded Department of Entomology, so she brought in several match boxes of scrupulously collected and labeled specimens for diagnosis and, hopefully, specific antiparasitic treatment. Donna even brought in her recently purchased microscope to demonstrate to Hal the corpses of parasites which she had collected from scrapings of her skin and scalp. Each specimen was labeled by date, time of day, site of collection, and so on. Hal accepted the cartons of specimens and calmly listened to her story of cumulative distress: personal, family, neighborhood, and church-related. This was not the highly intelligent, outgoing, and cheerful Donna that he knew in past years from annual church suppers. While he telephoned the specialist at Clemson assigned to extension-type consultations, she sat in his office, face drawn and tired from her burden of months of sleepless nights and frustration. Donna tried to read any expression on Hal's face as he spoke with the specialist.

"I can't promise anything, Donna. Professor Ivy is busy with research on poultry lice for several large companies, but he promises to find time to do this examination for us. We were classmates on the basketball team."

Donna sighed unhappily and left with little sense of relief; her face reflected months of disappointment and few answers. She had already experienced a series of failures to receive help from her family doctor, the dermatologist, and the neurologist at the medical center.

SCENE TWO

The scene shifts to Professor Ivy's lab at Clemson.

"Bud," he says to his busy graduate student, who is combing the feathers of a tray of ducklings, "here's another one! These matchboxes come here about once a year."

"How's that?" Bud asks.

"Hal, a really conscientious and dedicated guy in Dale County is asking for our help, and we can't say no. He has one of those 'fruitcakes' with her collection of matchboxes. They are always full of particles of skin, hair follicles, and carpet-fibers with nary a larval insect or adult insect part in sight."

"What should we do to help?" Bud asked.

Ivy replied, "I suppose the usual routine that my professor taught me, years ago, when I was his assistant. This time I am passing the job on to you. You will have to do the microscopy, document the negatives for our log book, send the results to Hal, and then brace yourself for the client's stressed-out phone calls. And believe me, she will call, and call, and call. Sorry, son, but it's your turn to learn how to handle it. This is a referral we dread at the university. Clemson seems to be the last hope for these people. They all wind up saying, 'Please, Professor, tell me I'm not crazy like everyone thinks!'"

SCENE THREE

"I can't thank you enough, Dr. French, nor can my husband, or my sons, or my friends, or the church. It's like I'm a new person! I can sleep a whole night again; the itching is gone. I don't have to spend all that time and money on carpet cleaners and pest control and referrals to specialists who don't help me.

"Do you realize that the last specialist, in neurology, insisted that I have had over six months of psychoneurodermatitis and should be receiving intensive psychotherapy? And intensive medication, since nothing else has worked?"

"Donna, you are not the first patient, nor the last, that I see with this really trying condition. It's been around for over one hundred years, first described in detail in Paris by a fine clinician named Thibierge. It's too bad that you didn't get diagnosed and treated earlier. You responded to treatment as well as, or better than, any of the patients that I have seen, so far."

"Thanks to Clemson and to ag med at MUSC! They told me to drive down right away to see you," Donna said. "Whenever I see any of my friends at the church or hear of cases in the community with my condition, I will make sure they get the right kind of help right away. When can I stop taking the Pimozide, doctor?"

"In another four to six weeks, Donna. I'll work with your local family doctor on refills if the condition should return, but that's not very likely."

"I can't thank you enough, doctor, bless you!" she said, shaking his hand warmly.

"I'm sorry, Donna, that we can't give you a cause-and-effect medical explanation for this syndrome, but fortunately, we have a treatment that really works for over ninety percent of cases, both here and in Europe. Unfortunately, it's so rare that the average primary care doctor doesn't see enough cases to recognize and treat it. Most patients go on and on without a diagnosis for an average of six months. Then they are sent to a psychiatrist as a last resort."

"I know, I know," Donna added. "But I knew I wasn't crazy!"

SCENE FOUR

"This slide, gentlemen, documents the clinical history of so called delusional parasitosis, as first recognized in 1894 in Paris, and confused with dementia or psychosis."

So began my brief slide talk for the Regional Entomology Society meeting in Charleston. The audience consisted of entomology experts and graduate students who might encounter one or two parasitosis patients in their teaching, extension service, or research careers.

During the discussion period, a senior professor rose to tell the group (with an inappropriate sense of humor) of his experience. His professor taught him how to handle matchbox clients with a "sure-fire" remedy: make up a Latin or Greek name for some imaginary parasite, so the client can "impress" their friends and doctors.

The good news, however, reached all of the younger entomologists in the lecture room. "Be empathetic, tell the truth," I urged them. "Refer them immediately to a medical center that knows of the latest treatment for this very debilitating disorder. Chronic insomnia and rejection by family and friends, due to constant itching and scratching, rejection by physicians who say 'this is out of my field'—this sad story is repeated again and again by these patients. The syndrome seems to prefer middle-aged, well-educated women, already wrestling with pre- and post-menopausal, life-change symptoms."

Summary: Why and How Does Pimozide Work?

No one knows, exactly. Designed as an antipsychotic agent, relatively safe with occasional arrhythmia side-effects, it was discovered during the sixties at the height of the drug craze in San Francisco. It proved to be the most effective treatment for the terrible symptoms of addicts, during withdrawal (things crawling under the skin). How it can interrupt the sensory cycle of nerve endings in the skin that trigger the relentless itching sensation has not been elucidated, so far. This is another example of serendipitous therapeutics. Once in a while, successful treatment precedes pathogenetic understanding of the clinical disorder, much like chincona bark. Also know as the "Jesuit's cure," chincona bark worked against malarial fevers centuries before the parasite in the blood cells was discovered by microscopists or quinine was isolated from the bark. Even medicine men applied poultices to wounds with moldy moss with some success, before penicillin and streptomycin were discovered.

The hallmark of success for this, fortunately, rare agromedical condition is earlier diagnosis, supportive care for the victim, and prompt treatment. In our South Carolina experience, the average delay in diagnosis and treatment was reduced from six to eight months to no longer than three months of debilitating symptoms. A seasoned entomologist at the University of Mississippi Medical Center has accumulated a series of referrals over the years, with similar delays in diagnosis but eventual successful treatment by alert dermatologists.

SCENE FIVE

True Parasitosis: My Son Is Not Crazy!

"What the *hell* is this!?" the surgeon, Dr. Jordan, snorted through his mask in the OR. "This is some kind of decay. What a foul mess!"

The OR nurse agreed, saying, "Yuck! I hope I don't throw up. Here are the tubes for microbiology."

Jordan barked, "They better test for all available antibiotic sensitivities, nurse." He suctioned the pus from the single large chest wound, trimmed the edges, irrigated with antibiotics, and packed the wound with a large pressure dressing. Helping push the patient on the gurney to the recovery room, the surgeon wondered what to do next. *This crazy eighteen-year-old with his overprotective mom,* he thought. *This is really something else! What a cock-n-bull story of waking in the night, feeling something crawling inside his chest, screaming and seeing flying things come out and fly toward the ceiling light. I'm sure his hysterical mom was no help to calm him down. I wonder what triggered the poor guy's complete breakdown. He must have been okay to pass the Army Reserve physical. I'll call psychiatry for a consult stat, and order a drug screen before we send him home. They better get him in detox before he hurts himself or someone else.*

SCENE SIX

A week later, our agromedicine office received a call from the mother, reciting the story of her son's strange wound, his hospitalization, and his vivid, strange experience requiring surgery. "How can we help?" I asked her.

"Please help me explain to the doctors here that my son, Johnnie, is *not* crazy. Surely you know what could cause all of this."

"I'm not sure, ma'am," I admitted, "but we can certainly try."

SCENE SEVEN

Using our university library and a MED-LINE search, I was coming up with a possible explanation for the unusual history of the young Army reservist. Two weeks prior to the fateful night of nightmares and flying things (blue-green flies swarming toward the ceiling light), Johnnie was out training with his company on strenuous all day marches with heavy back-pack equipment to toughen up for the regular Army. The heat and humidity along the marshy swamps of the Savannah River forced the young men to strip to the waist, even though their pack strap and buckles rubbed their chest muscles raw. At night, in the simulated jungle of the swampy marshland, some used mosquito nets and repellants, and some didn't. The heat and exhaustion seemed worse than the gnats.

Our research kept coming up with *myiasis,* a kind of botfly infestation of wounds, well known in tropical medicine, especially in Latin America, Africa, and Southeast Asia. Case studies in the literature made Johnnie's case sound almost benign. Adult, blue-green flies tend to lay their eggs in small wounds around the eyes, ears, and nasal passages in sleeping victims. Weeks later, boils appear under the skin or mucous membranes. The eggs mature into larva (hungry crawlers eager to digest living tissues), until as pupae they can penetrate to the surface (skin or warm-blooded animal hide) and emerge to fly free and begin the life cycle again. Gruesome case studies documented invasion of the facial sinuses, the mastoid, even the brain!

On the other hand, the literature noted that maggots were used medicinally and surgically to clean up nasty wounds and ulcers, neatly digesting away the damaged or less viable tissue, to permit better healing of wounds. In World War I, surgical dressings with maggots were applied to some battlefield gangrenous wounds with some remarkable benefit (before the age of antibiotics).

SCENE EIGHT

Clemson entomology professor, Dr. Kent, answered my call for his help on the subject of botflies, maggots, and the like.

"Stan," Dr. Kent said, "I'm not at all surprised by the unfortunate soldier's story. The surgeon just needed to get some professional Clemson help. We would have told him to put some of the infested tissue in a bottle, and we could have identified some of the hundreds of eggs and larva in the abscess. Once they get going, the abscess can leave a nasty wound. It's a well-known problem for livestock, and veterinarians say the abscesses may be resistant to many antibiotics, due to secondary infection.

"I have a grad student here this fall, and he will gladly make a home-visit. He can search the boy's bedroom for any residual blue-green flies, adults or pupae casings. The casings resemble rat droppings or cockroach egg cases.

"Incidentally, my retired professor friend has such a deep interest in *Calliphoridae* that he is in demand to help crime labs all over the country. In the U.S. the blowfly is a valuable forensic tool, because its pattern of colonizing and digesting decomposing flesh is like a time clock, quite helpful in estimating the most likely time of death.

"In most of the United States, *Calliphoridae* are attracted to a corpse within minutes after death. When I was in Guatemala, Stan, I learned from the natives working the lowlands, that any suspicious itching lump under the skin could be an infestation. Since most peasants chew tobacco, they simply treat the area with a generous dose of tobacco juice. It dries up, closing the central respiratory pore of any single moving larva. Plus, tobacco is a good insecticide. Ordinary Vaseline applied to the pores would work as well, I think."

POSTSCRIPT

In case one thinks that myiasis is unlikely to occur outside southeastern United States (where cattle screw-worm infestation had to be eradicated by sterile male fly treatments decades ago), think again. Although *Calliphoridae* cannot withstand frost, they can be imported. Consider the alert diagnosis and treatment of a sixty-five-year-old man at the Brigham and Women's Hospital in Boston, MA. According to the June 9, 2005 report in the *New England Journal of Medicine,* six weeks after returning from a vacation in Belize at the beach and touring the rain forest, the patient developed five nodules of various sizes on his chest and left arm. The lesions occasionally hurt, were inflamed, and had a central pore, through which a dark, foul-smelling exudate drained.

Despite two weeks of antibiotic treatment, the lesions would not heal until each of the five nodules yielded *Dermatobia hominis* (human botfly) larvae, extracted with manual pressure and tweezers under local anesthesia. They are distributed throughout Latin America and will be attracted to any skin wound, body heat, and the odor of decaying flesh. Teamwork at the Boston Medical Center between the internist, the surgeon, and an entomologist provided the diagnostic and therapeutic knowledge required for specific diagnosis and treatment of this healthy traveler, who was trying to cope with this unusual kind of "enemy within."

REFERENCE

Marty FM, Whiteside KR. Myiasis due to *Dermatobia hominis* (Human Botfly). *N Engl J Med* 2005;352:e21.

Chapter 7

Infertility: The Farm Wife, the In-Laws, and the Physician

SCENE ONE

What a great day! I thought, as young Dr. Lee greeted me warmly in the driveway of a strikingly large and beautifully landscaped dairy farm. He had been one of my favorite family medicine residents. Now he was a very successful hometown boy who excelled in private practice and was honored as Physician of the Year in the county.

It was spring, azaleas blooming and birdlife busy, as our three cars approached a smiling, welcoming six-footer, sitting high on a huge new John Deere tractor. "Good morning, y'all! Welcome to our farm. Looks like you brought the whole crew," he said as he watched the camera crew begin to unload their equipment from their university van. Smitty had kindly consented to participate in an ag-med/preventive medicine video designed for medical students to acquaint them with current farm hazards and farm safety procedures on one of the state's largest and best known Guernsey dairy operations.

Rainbows in Washtubs

doi:10.1300/5851_07

Luck was with us for a successful filming by the crew. The weather and, most importantly, the goodwill of Dr. Lee and Smitty, who had been classmates at Clemson University not too long ago, were on our side. In no time, the crew captured excellent shots of the hazards of power-take-off equipment and the importance of using protective shields to prevent serious injury. In addition, the roll-over bar on the new tractor was filmed to point out the all-too-frequent crushing injuries to the driver. The need for hearing protection was made obvious as the power machinery got revved up to full speed. On camera, Smitty was given a broad-brimmed hat by Dr. Lee, warning him to protect his fair skin from UV light. He told Smitty to save his less protective baseball cap for casual wear or coaching Little League games.

As the film crew packed up their gear, giving Sparky the collie a fond farewell, they were all smiles. They were sure they had good footage for a first-rate, fifteen minute agromedicine video. Having Dr. Lee and Smitty interacting so easily could hardly have been scripted as well, and actors would have been hard pressed to be so at ease with each other. The dairy farm setting was an ideal background and would appeal to medical students who may be thinking of going into small town practice some day.

So much for cinematics.

I noted Smitty and Dr. Lee engaged in earnest conversation, well out of my earshot. As I ambled over to play with Sparky and his bone toy, I noted a young woman in broad-brimmed sun hat, colorful blouse, and blue jeans coming down the path from a farmhouse (one of two built side by side). She was headed to the single large rural mailbox located at the foot of the common driveway. "Good morning, I'm Alice," she smiled, "you must be the doctor from MUSC that Dr. Lee said was coming. Some nice excitement for us country folks."

"Good morning to you, ma'am," I said. "Thank you all for letting us visit. We should have some worthwhile video to share with the med students and nursing students. Some will surely want to practice in beautiful rural areas like this." The collie greeted her enthusiastically.

"You know, doctor, the three of us—Smitty, Dr. Lee, and I—were all the best of friends at Clemson. When he was pre-med, Smitty was captain of the tennis team and I—well, I was vice president of the student council. Those were the days. Lots of fond memories, Doc married Maryanne, the secretary of the council. We had a great time double dating, having Bible study, and enjoying football weekends!"

Alice flipped open the huge mailbox, she sorted out the majority (catalogues, business mail, etc.) to replace in the box. A scant few pieces of mail were retained in her hand. Smiling again, Alice offered me some iced tea and led me to a swing on the porch of the smaller of the two farmhouses.

"Our Dr. Lee says that you do research in Charleston and teach family medicine."

"Yes, I do," I said, sipping the rich southern brew.

"Do you know anything about all these chemicals we have to use on the farm? You know, to control the weeds in pasture and horseflies in the milking barn?" Her look was suddenly serious and intense.

"A little," I answered, trying to appear modest. Then there was a long pause. I knew she was leading up to another searching question.

Looking at me intently, she said, "My brother-in-law, George, is a PhD in Baltimore doing research on reproduction in a cattle-breeding operation. He seems convinced that our problem—our problem having a baby—is probably due to Smitty's overexposure to herbicides, insecticides, and other chemicals being used on the farm.

"He says that's one personal health reason he left Papa, Smitty, and the whole dairy business to start a different life in veterinary research. As the oldest son, George Jr. should have taken over the farm from Papa instead of Smitty. Even though my husband and I were enjoying living and working in Hilton Head, my father-in-law insisted we return here to carry on the family tradition. Smitty loved working as a tennis pro, and I was being promoted to a department store manager. But hardly anybody says no to Papa. Papa and Mawmaw are statewide leaders in agriculture and the Farm Bureau. They are so proud of their traditions and maintaining this valuable farm land and prize herds. They count on passing all this on to their grandchildren and great grands."

Alice turned her saddened face away from me. Listening to her story on the porch swing, I suspected from the quaver in her voice that she had to stop talking about having children to wipe her eyes. Even as a physician, I felt embarrassed. I was an intruder in a very personal and private matter, which she couldn't help but bring up. I looked around for Dr. Lee. From the earnestness of his expression, I suspected he was hearing the same infertility story from Smitty.

Then Alice's expression hardened, as she crushed a piece of junk mail in her free hand. "You know, Doc, out here—I'm from Atlanta, you know—out here, I can't even have my own mailbox! Mawmaw likes to get the mail first thing every morning and hand me my pieces like she was doing me a favor. I just don't know what we can do to have a baby and get on with our lives!" Another tear formed in her eye. "Maybe Papa will let Smitty go—let *us* go—back to Hilton Head and Smitty's own career!"

SCENE TWO

Leaving the farm with Dr. Lee to pick my car up at his office, I detected a tear in his eye. "What a fine couple, what a shame, Stan. They need to live lives of their own. Papa, you see, is not only a proud dictator; he's also in need of serious medication. Bipolar depression runs in his family, and he's ashamed of needing any kind of medical help or admitting any weakness on his part."

Heading back to Charleston, alone in my car, I had much to ponder. I needed to update my files on reproductive side effects from occupational exposures, especially in dairy farming, as well as the effect of emotional stress on fertility. Second, I empathized with Dr. Lee and his wife, Maryanne, happily expecting their fourth child, while Smitty and Alice were still trying for their first. Third, I pondered the dynamics of generational stress on the farm, when honoring and sustaining the land is so important to the family. This burden for the children and grandchildren can prove to be a blessing or a curse. The romantic picture of happy American farm families keeping up the frontier rural values from generation to generation may not be as valid as one might think, or as timeless as one might hope.

Uncertain crops, weather, and economic pressures such as fickle markets, tariffs, and the rising costs of machinery, bank loans, seeds, agrichemicals, fuel, and utilities (not to forget simple water!) can lead to a kind of depression and overwhelming loss of purpose in life, especially if the sense of proud tradition is unfulfilled. Suicide is not unknown in farm life, and physicians like Dr. Lee are the thin red line of defense and medical hope for earlier diagnosis and specific medication.

Chapter 8

Fire Ants, Ticks, and Spiders: Who Is at Risk?

One of the key first lessons drilled into the USPHS doctors, at the CDC in Atlanta, aspiring to become epidemiologists is to ask the following questions:

How to define a case of X, Y, or Z?
How to define a noncase?
How to define the kind of exposure?
The level of exposure?
The timing of the exposure?

Does this sound familiar? It should (see Chapter 2, "Tools of Discovery"). They are simple tools, but in the hands of an expert, a craftsman, the results will be impressive and quite illuminating. Recall Figure 2.2. Is the case sicker than the noncase? Is the likely exposure present or absent? The answers yes/yes define an exposed case and no/no define a nonexposed, noncase, a true "control." In the case of frequent exposure to the unfriendly world of biting, stinging, and feeding insects we humans can become true cases. However, the nature of our "illness" or responses to their bites, stings, or feeding practices can vary. There is a spectrum of clinical reactions to insect venom, for example, ranging from trivial to life threatening (see Table 8.1).

Rainbows in Washtubs

doi:10.1300/5851_08

The following stories of insect-related agromedical discovery will underscore the value of the epidemiological approach shown in Figure 2.2 to what may be a new or emerging health problem, as a new type of insect invades a region and challenges the diagnostic skills of health care providers.

SCENE ONE

Dr. Martin had let his nurse off early for her son's soccer game. Alone in his office, he was turning off his computer, picking up the children's storybooks in the waiting room, and adjusting the thermostat for the night, when he heard a *thump! thump!* on his entrance door. *Who could that be,* he thought, *well after office hours?* Dr. Martin was not expecting anyone he had consulted during any of his late phone calls. *Thump!* And *thump!* and then a very diminished *knock.* Dr. Martin rushed to unlock the door. As soon as he opened the door, he stepped back at the sight of the body of a young man slumped in a heap across the threshold. When he rolled the body over he was stunned to see a desperate looking, ashen-faced youngster pointing to his throat, gasping for breath. Dragging him into his foyer, Dr. Martin quickly noted the puffy eyes, the swollen lips, and the strident wheezing from his chest and raced to the refrigerator for one of his Epinephrine spring-loaded injector pens. Precious minutes after injection, Dr. Martin could see the man begin to sit up and the panic fade from his face, as his breathing became easier.

Thank God! Dr. Martin thought. *I'm not sure I could have intubated him without some help.*

A clear and simple case of anaphylactic shock with near-cardiac arrest, just minutes or seconds away from sudden death.

As the confused stranger tried to sit up, he sliped off his heavy backpack, rubbed his eyes, and stared at Dr. Martin. He took long, deep, satisfying breaths. While waiting for him to speak, Dr. Mar-

TABLE 8.1. Insect-Human Interactions That Trigger Agromedical Investigation

Insect	Reaction	At risk (who and where)	Diseases	Treatment	Prevention	Epidemiology
Fire ant	Allergy is unpredictable (1-2 % of population)	Homeowners Infants Children Gardeners Landscapers	3 responses: mild (75-80%) moderate (16%) severe (1-2%)	Epinephrine Desensitize Avoidance	Avoidance Early diagnosis critical for 1-2%	Imported ant needs Latin American competitors in the U.S.
Tick	Infection by microbes	Campers Hunters Hikers Surveyors	RMSF Lyme Ehrlich	Early doxycycline for all 3	Avoidance Remove ticks	Lyme vaccine for dogs
Spider	Necrosis by brown recluse species	Electricians Carpenters House cleaner	2% infection	Rest, elevate Oral anti-biotic	Avoidance	Geographic Spread across U.S.
"Puss" caterpillar	Toxin	Outdoors	Local pain; severe	Scotch tape removal Ice Alcohol Analgesic	Avoidance	Little research, so far
Bees	Allergy "Killer" bee attacks	Outdoors	Anaphylaxis Massive toxic stings	Eninephrine Desensitize Avoidance	Avoidance	Early diagnosis Economics of bees essential to agriculture
Cockroach	Disgusting infestation	Asthmatics 25 % of children	Nondisease agent Asthma a concern	None	Garbage & moisture control	Sanitation, housing, & moisture control
Mosquito	Infection	Outdoors	W. nile virus Encephalitis Malaria/yellow fever	Supportive Antimalarials	Mosquito abatement	Tropical diseases Travel hazard Malaria vaccine(?)
Fleas	Allergy	Indoors Pet owners	Dermatitis	Avoidance	Pet care	Carpets Allergy to dust Bedding

tin thought this poor kid could have been wiped out, as surely, as if he had been hit by an eighteen-wheeler on the highway. The colorful stickers on his hiking bag carried labels from state parks from Vermont, Massachusetts, Virginia, and North Carolina. His loaded bike was in a heap outside the doctor's door. Later, piecing the man's story together, Dr. Martin learned that he left home in Maine ten days earlier. The man, Jack, had made great time on his solitary journey, cycling successfully without incident, until he stopped to rest with a sandwich at a roadside rest stop. While sitting against his pack, he swatted a few gnats and felt a stinging sensation, several in fact, above his high top boot on his left foot. Looking down, he brushed away a few small ants, brownish-red, not unlike the friendly picnic-visiting ants back home in Maine. Within minutes, Jack began to feel itching places around his neck, cheeks, and scalp. *Must be these pesky Carolina gnats,* he had thought and applied some rubbing alcohol from his emergency kit.

He knew he wasn't allergic like his younger sister who got hives from bee stings, and the cooling alcohol seemed to help. He was only a few miles from the suburb of Charleston, where his college friend lived, and within earshot of the glorious surf awaiting him at the Isle of Palms.

Dr. Martin felt sorry for the young man, as he got no answer from phoning his friend's house. He decided to take Jack home with him and provide him with supper, a hot shower, and a place to rest; Dr. Martin also wanted a chance to observe any allergic relapse during the night.

Later, Dr. Martin related the incident at our teaching rounds for agromedicine. Soon, the whole story of the red imported fire ant invasion from Latin America would begin to unfold, challenging doctors, entomologists, and farmers alike. Without any effective controls and with no natural predators, the *Solenopsis invicta* species would spread throughout the Southeast, from Alabama and Florida to Texas and beyond—wherever potted plants and produce were transported, and where severe frost conditions do not regularly occur.

Solenopsis invicta, most appropriately named, hitchhiked its way on a tramp steamer from Argentina to the port of Mobile, Alabama, in the early 1930s. An unusually aggressive stinger, the RIFA worker ants defend the home subterranean nest and their precious queen against any semblance (vibration or odor or threatening movement) of an invader. At the slightest provocation, worker ants swarm, to aggressively "boil" out of multiple channels to attack the intruder. By comparison, ordinary native U.S. carpenter ants and picnic ants are like domestic pets, charming and timid, compared to these ferocious critters. RIFA not only attack humans, they attack newborn lambs, calves, and ferociously take over the territory of any nearby non–RIFA colony. Thus, they have steadily spread from Mobile to all of the Southeastern states, especially the warmer sandy soil of coastal counties. The only apparent limiting factor appears to be nighttime or seasonal colder temperatures, when they retreat into their extensive network of subterranean channels and nests.

Research into RIFA has been quite limited and funding, by any agency, has been slow to surface. (They have not yet invaded the White House grounds for the Easter egg hunt.) If they could be indicted as a legitimate threat to crops or livestock, the land-grant campuses would be more motivated to study them. In fact, Louisiana rice and sugarcane growers claim RIFA are beneficial. They seem to attack some of their economically damaging insect pests. Not so in most of the Sun Belt states, from the East coast to the West coast, where California is energetically funding RIFA research to track and deter the intruder.

The turning point for RIFA survival and territorial expansion throughout the Sun Belt was not just importation to seaports from Latin America, but the absence of RIFA's natural enemies that exist in its native habitat: the phorid fly, which feeds on them, and other competitive ant species whose colonies repel the aggressive

RIFA so effectively that they never number more than 5 to 10 percent of the ant population in South America.

Returning to the strange case of the near-fatal sting of the Maine hitchhiker, Dr. Martin and his colleagues in the specialty of allergy had much to ponder over the incident. Questions needed answering. For example, how could this nonatopic teenager (no previous asthma or food allergies) react so violently to a few ant stings?

The answer is that RIFA anaphylaxis only affects 1 to 2 percent of the general population, *regardless* of any genetic predisposition to allergens (nonatopics and atopics alike). How could Jack develop RIFA hypersensitivity so quickly? It was his first trip below the Mason-Dixon Line. His bike trip was quite swift, less than five days from northern Virginia to RIFA–infested South Carolina: Normally the sero-immune (IgA) mediated response takes three to four weeks to develop specific antivenom antibodies.

After Dr. Martin's case presentation at grand rounds, I was motivated to learn more about this strange imported ant. Was it merely another garden or woodland nuisance, or was it a growing cause of morbidity among unsuspecting patients and a challenge for their medical caretakers in our state?

Responses to my early inquiries ranged from complete denial by a well-respected Clemson extension agent ("they don't exist in the Upcountry, our foothills of Appalachia"), to enterprising Charleston pharmacists selling a goodly amount of antibiotics and steroid creams and ointments for any RIFA blister-like sting on any customer showing a pustular rash. Actually, the classic typical RIFA pustular-looking bleb (a standard response to a single venomous sting) is purely chemotactic—a sterile bleb with no infectious consequences, unlike true impetigo. Characteristically, the pearl-like blebs will be clustered in a small semi-circle of skin where the diligent worker-soldier ant has injected several doses of venom from his posterior stinger. Unlike the honey bee, the RIFA ant does not eviscerate and die after stinging but can "reload" venom and sting again with a permanent stinger. For the patient, it is important to avoid scratching the itching, burning blebs, while awake or asleep, because a broken bleb allows skin bacteria (streptococcal and staphylococcal) to colonize the area.

Are fire ants a nuisance like most other ants? I thought it over again, and as a trained epidemiologist, I found the first case of Jack either an anomaly or a clear signal to start the classic medical detective process. I would have to ask, step by step, a simple series of epidemiologic questions:

1. Is this case of X unique or capable of happening to others?
2. If it is not unique, how many similar cases are there? Be sure to define a case with strict criteria. No room here for fuzzy thinking. No "near" cases or "could-be" cases to be counted. They would only cloud the picture.
3. If it is poorly recognized by doctors and patients, how can early diagnosis and reporting be improved?
4. How do we confirm diagnosis? Identify clinically or microbially?
5. What is the spectrum of clinical response to the agent?
6. What are the exposure characteristics that separate cases from noncases?
7. What are the clues to prevention or control of the specific agent or exposure? (In the case of ticks, long sleeves, boots, light clothing, tick repellents, shampoo and shower removal of unattached ticks ready to attach and feed.) Avoid grassy infested areas, vaccinate dogs, and practice tick hygiene for cats and dogs.
8. How do we evaluate any progress in reducing the number of cases of RIFA anaphylaxis in our state?

As time went on, our staff developed a brief questionnaire and mailed it to all the registered licensed primary care and ER physicians in the state. Meanwhile, more phone calls concerning incidents of RIFA stings and near-fatal reactions kept coming into our agromedicine office from patients, emergency room personnel, and extension offices around the state. The health department offices referred their calls to Clemson Extension offices, since such stings were not "a reportable disease, or public health problem."

Consider the following incidents of RIFA-triggered allergic reactions that came to our agromedicine office:

- An avid gardener who moved from Canada to the Low Country recently can no longer tend her yard. She is having more and more alarming symptoms from RIFA stings coming from visible mounds and invisible underground channels. She is ready to return to Canada.
- An ER physician treats a rural resident twice for near-fatal anaphylaxis. He wisely will not discharge her from the hospital until she has a qualified exterminator treat the RIFA mounds located under her bedroom in the trailer home. Multiple nests of RIFA ants were crawling up through the cracks in the patient's floorboards and cruising along the electrical wiring channels. RIFA are strangely attracted to electrical or magnetic fields—more research topics for graduate students in entomology.
- A toddler trips in his back yard near the air conditioner, falling face down into a fire ant mound. Dozens of worker ants attack his lips, nostrils, and throat. A 911 call barely gets him to the ER in time.
- An MD rushes his wife to the ER after she is stung getting into her car at an urban mall. The parking lot is paved concrete, but RIFA ants are active along the seams of asphalt and concrete. They seem to like the warmth such surfaces hold.
- A grandmother who came from the north to help her daughter is hanging diapers outdoors. She is rushed to the ER for acute respiratory distress. She must have been sensitized to the RIFA venom during the previous visits, but her north-

eastern doctors had no diagnostic experience to explain her allergic reactions when she returned home.

- A bed-ridden patient in a second floor room of a centrally located, urban hospital is stung repeatedly by RIFA swarming from the window unit air conditioner. The ants are attracted by visitors' fruit baskets.
- A nursing home resident is found dead, slumped in a rocking chair on the front porch, covered with fire ants, with too many stings to count. The coroner is puzzled, not able to time the RIFA stings: before or after the final fatal moments? The family of the deceased is outraged and hires an attorney to sue for negligence. Is anaphylaxis or stroke the cause of death?
- A healthy golfer is attacked on the ankle and collapses suddenly. His death is diagnosed in the ER as acute coronary attack, with no pre-mortem evidence of any skin manifestation of RIFA stings. He probably died before any local reaction could become visible. Another sudden death from latent coronary disease or fatal anaphylaxis?
- Hoping to avoid aggressive RIFA stings, EMS workers now carry insecticide sprays and plastic sheets to protect themselves when attending traffic casualties at any roadside crash. Fire ant mounds are now lining roadsides along all major South Carolina highways, disseminated by maintenance crews with lawn and weed cutters.
- A three-month-old infant is rushed to the ER after suffering multiple sting blisters on her face and forehead while in her crib. Surprisingly she has no other symptoms and is discharged from the nursery after thirty-six hours. A color photo documents multiple facial lesions (pustule-like blebs) and periorbital swelling. How the fire ants crawled into the crib (foraging for bits of baby food or dried formula on the baby's cheeks) was never explained to the doctors.

Our agromedicine office mailed out the statewide survey, mentioned previously, on insect-related morbidity. There was a remarkable 70 percent response, indicating a willingness to help understand this spreading problem.

The doctors documented a spectrum of RIFA morbidity in their practices, ranging from telephone calls, to office visits with secondary infection of stings, to emergency hospitalization. The survey also revealed a lack of proper referral by primary care providers of anaphylactoid patients to allergists for accurate diagnosis and effective desensitization. Referral is critical for the 1 to 2 percent highly susceptible minority of cases who will experience severe hyperimmune reactions to RIFA venom (no matter if it is a single ant's sting), unless they are desensitized by the allergy specialist.

SCENE TWO

"This can be a lonely place," I said to Dr. North, as we were seated side-by-side at the annual Scientific Session in Atlanta. Our job was to stand by our exhibit posters. His was titled "Physician's Guide to Arthropods of Medical Importance," representing his southern medical center. Mine was titled "Physician Survey of Fire Ant, Tick and Spider Bites," from the Medical University of South Carolina.

"I agree, Stan," he said, offering me a stick of gum. "Maybe we should be in a better location, near the rest rooms or near the free refreshments."

It was a long afternoon, babysitting our scientific posters and our carefully prepared handouts of useful medical information. "Jerry, maybe we need to bring free pen and pencil sets, or golf-gloves, or tennis balls to lure them," I joked, "like the well-equipped pharmaceutical representative."

"Fat chance, Stan, on our slim budgets!" Feeling sorry for ourselves, we welcomed a few medical visitors who wandered by, wanting to share their stories of arthropod-related illnesses, encountered in gardening, hunting, hiking, home repairs, or clearing brush around their homes. Each visitor to our exhibits commented, "I wish we had this information in med school," or "If only my child's reaction was diagnosed, it could have been treated earlier" (brown recluse spider bite).

Over dinner that night, I wished Jerry success on his new book and assured him that we would use it diligently with our students and residents in Charleston.

"You know, Stan," he confided, "you would be surprised how many medical calls I get at the health department, as well as insect specimens to diagnose. There's a real need for some medical entomology resource to help the public in every region, maybe every state."

"I wish you were in Charleston," I say. "It would relieve me and some of my colleagues at Clemson of a lot of puzzling signs and symptoms to account for."

"I wish you were in at my university, Stan. The folks at our med school don't really appreciate what we can do for them. That's why I had to come up with this book. Thanks to the publisher, it might help some doctors who want to make a better diagnosis."

"Was it always like this, Jerry . . . the gap between entomology and medicine?"

"Not in tropical medicine, for sure: malaria, yellow fever, sleeping sickness, encephalitis, elephantiasis, encephalitis, dengue, hemorrhagic fever, plague. And not when I served in World War II. Without DDT our troops in Italy would be unable to fight the Germans, because of louse-borne typhus—no immunity in Americans."

"So," I replied, "in wartime, the entomologist is a vital part of the team. But in peacetime, Jerry, it seems that your kind of specialist has to find work in agriculture, the pesticide industry, a large public health division, or teaching biologic science in a university."

"That's the way it is, I guess, Stan, until the next global war, heaven forbid!"

POSTSCRIPT

The "good" news is that at least one segment of our society has a growing interest in medical entomology: liability attorneys. *USA Today* (3/8/2006) reports a 20 million dollar suit against a Catskills (New York) resort. After three nights at the 700-room hotel, the claimant had over a dozen color photos to document some 500 bedbug bites on her back, chest, and legs. Several other lawsuits have been filed around the country for alleged bedbug attacks in hotels. Primary care physicians may have to keep up with attorneys learning about health effects of arthropods.

Table 8.1 lists some of the health-related effects of insects that perennially raise questions from our clients in the Agromedicine Program.

Chapter 9

A Farmer's Risks: How to React?

SCENE ONE

O'Neal loves his dairy farm, but his life is not for everyone. Dairy farming has a high dropout rate. Long days, long hours, and an unforgiving routine don't allow much free time, much less a vacation, unless you can find someone to relieve you for a week or two. O'Neal's vocation includes improving his herd, detecting vet problems before they get out of hand, dealing with equipment repairs and replacement, and worst of all, keeping up with the market. Fortunately, his wife, Sara, has been a perfect helpmate, lending a hand with the milking, keeping the books, and still working part-time in town at the bank. She's always been good with figures. In a few years, ten or so, he might think about turning the dairy over to his loyal and steady helper, his sister's son. Then something he didn't foresee or expect took over his life—cancer!

It all started with a little loss of hearing in his right ear. He thought it was wax build up. The pharmacist recommended some ear drops, which helped a little. But the stuffy feeling got worse, and he noted some swelling in his sun-tanned neck, just below the ear. Sara insisted he see the family doctor.

Rainbows in Washtubs

doi:10.1300/5851_09

The next thing O'Neal knew, he was in the care of the big city surgeon and was being told that the lymph node biopsy was "not good, not good at all." He would need a radical neck dissection for a slow growing but extensive carcinoma (basal cell), which should have been caught early. It started as an ulcerated lesion which failed to heal (a "rodent ulcer," like a rat chewing on your skin). It began on his right ear tip where the hours and days of unprotected exposure to sunlight had free rein on his fair, Scotch-Irish skin. *If only he had come earlier,* the surgeon thought to himself. O'Neal's prognosis was not good—fifty-fifty for metastases, even with post-op chemotherapy.

When I made a farm-site visit with medical students and O'Neal's long-time family doctor, guess what? O'Neal was still wearing his favorite green John Deere baseball cap. Protection again UV light still comes to him as an afterthought, not a priority, even after cancer surgery.

There was much to ponder with Dr. Lee, his kind and caring family physician. Maybe we can get the word out to his farming patients. Maybe they will start wearing the broad-brimmed hats and bandana-neckerchiefs like their granddads wore, protecting them against the heat and sunburn. Dr. Lee agreed; he would speak at the next dairy producers' regional meeting.

SCENE TWO

Consider the range and intensity of risks experienced on an average farm: the weather (including drought, flooding, tornadoes, lightning, and frost), livestock risks and care, crop failures, pests and blights, machinery aging and breakdown, tractor roll-over, power-take-off injuries, agrichemicals in concentrated strengths, dangerous liquid ammonia applications as fertilizer, ground spray of chemicals, premature hearing loss and UV-aggravated cataracts, skin cancers, entrapment in silos and asphyxiation in granaries, farmer's lung from overexposure to hay microbes, fungi, dust and silica from dust, isolation from human contact during work for long hours and days, trauma from animals (crushing injuries, bites), heat stroke, taxes, loans, foreclosures, failing markets, and lawsuits from neighbors who resent animal waste odors and toxic chemical drift over their new subdivision homes. Then there is the threat of government takeover of wild life habitat adjoining farm land, in the name of conservation.

In farming, how can one learn to react, but not over react, to potential health threats? How can O'Neal, and farmers like him, learn new habits to protect themselves? In agromedicine, the health promoting goal is to learn and disseminate better ways to estimate risk to avoid a variety of acute and chronic hazards of the agricultural way of life. The exposure to a variety of insect bites and stings and transmission of disease, as described in Chapter 8, represents a typical agromedical challenge.

In South Carolina, in the late 1960s and early 1970s, pioneer physicians responded to the call from the EPA to set up a laboratory to evaluate human exposure to increasing use of organochlorine pesticides (including DDT and its successors). Organophosphates and the newer herbicides and fungicides demanded further medical research, which lead to field studies of worker exposure and advances in laboratory measurement in parts per million, and then parts per billion.

When I joined the SC Pesticide Human Health Effects Laboratory in the early 1980s the focus of interest and finding was still biochemical and toxicological. As an epidemiologically trained clinician, I found plenty of challenge to broaden the mission of preventive medicine for farmers and their families. Some of the challenges have been suggested in the other narratives in this book—from insect anaphylaxis to food safety issues—all requiring intensification of MUSC–Clemson interdisciplinary teamwork. But most impressive to me, as an associate physician in the EPA program, was the opportunity for a strategic, practical reevaluation of the variety of nontoxicologic risks experienced by our clients in the forty-six semirural counties of South Carolina. This wider perspective on agricultural health and safety only became evident and feasible for me as I traveled around our state. There was no substitute for delivering consultations, going to community and commodity meetings, farm visits, and visiting rural physicians and nurses. My colleagues and I managed to travel in state over 1,500 miles a month. Over time, our small team evolved to represent the medical prevention arm of Clemson's research and extension network.

What did we learn from our grass roots travel and observation? First, premature hearing loss was prevalent in many middle-aged farmers due to a lack of hearing protection when using power machinery. Second, sun-related skin cancers and precancers were occurring increasingly and needlessly. Third, missing fingers and scars from trauma on the farm were too frequent among farm workers and their helping children. But overall, a strong work ethic and a culture of cheerful, optimistic denial of the hazards of their agricultural calling persisted.

In addition, growing concerns were being voiced by the agriculture community over possible reproductive hazards, as well as hematalogic tumors (leukemia and lymphoma), allegedly connected to pesticide exposure. Rural physicians were increasingly concerned over hidden depression, alcoholism, child abuse, marital discord, neglected hypertension, obesity, and diabetes.

For me, the romantic picture of bucolic, American family life was begging for a more objective perspective. This led to programmatic reduction in emphasis on agrichemical poisoning (as the EPA and industry were working together to reduce the problem significantly). I came to appreciate ordinary issues such as:

- preventing heat stroke and exhaustion
- predisposition to injuries from farm machinery
- neglect of eye injuries and secondary infection

- maintaining tetanus immunization
- attention to geriatric risks for older farmers
- child injury from exposure to livestock and farm machinery
- stress from economics, overwork, marital discord, teenage rebellion, and generational pressure to "save the farm," as well as urban encroachment

One by one, these varied personal, environmental, and cultural issues all emerged as grist for our mill. Health promotion cannot succeed on the farm with tunnel vision and narrow interests, or a closed mind. To illustrate this agromedical approach, one can review the following list of educational brochure topics that our program provided to the forty-six county extension offices. The format was designed to be one page, front and back, with phone numbers (before e-mail) for free phone calls for assistance from MUSC (our office, forty available hours per week) and/or Clemson. Generally, no more than ten tips for prevention (questions and answers) were acceptable for this concise format.

Heat Stroke
Skin Cancer
Hearing Loss
Eye Injury
The Older Farmer
Breast Cancer and the Environment
Fire Ants
Florist Dermatitis
Food Safety
Diet and Cancer
Tickborne Disease
Child Safety
Trauma Prevention
Rabies Prevention
Alar and Apples: Fact and Fear
Chemophobia

Signs and Symptoms of Acute Pesticide Poisoning
Poisonous Mushrooms
The Community Pharmacist
Stress Management
Nitrates in Ground Water
Farmers Lung
Bio-Engineering—Genetically Modified Seeds
Food Terrorism
Tickborne Diseases
Plant Toxins and Allergens

SUMMARY

The enduring question of farming hazards and risk appraisal is, how much is enough? Then how does one react (to any given hazard or health risk) and not overreact? This is a personal question, a family question, a medical caretaker's question, a preventive medicine question, and a programmatic question for an inter-university agromedical public service program. The health questions seem ever-changing and the responses must be equally adaptive.

Food safety issues began decades ago with rural farm kitchen hygiene to large scale food production and distribution issues, involving microbes, chemical residues, and potential allergens in the food supply from farm to table.

Farm safety issues have to keep up with the changing technology of power tools and equipment. One of the major causes of preventable farm deaths and injury nationwide occurs on the highway in motor vehicles transporting vital supplies and laborers crops to market.

Chemical issues have evolved from aerial sprays and ground-rig sprays to integrated crop management techniques and bio-engineering of seeds, to ground water and air-drift issues. The programmatic implications of this ongoing interplay between hazards, potential hazards, and human health protection are not simple.

In summary, it can be briefly stated that the more flexible and adaptive the agromedical program, utilizing client-based research and client-based prevention, the more likely it will keep up with the times. In essence, the community-based rural practitioner of medicine and the extension service are the key sentinels for emerging problems and potential problems. The key to early surveillance of problems is attention to better listening, being on call for help, and networking with available scientific expertise at the university level.

Our dairy farmer with the inadequate baseball cap, O'Neal, may or may not recover from his neglected skin cancer. He may out live the surgeon's pessimistic prognosis. O'Neal may soon retire to a beachside community in Key West, enjoying deep sea fishing and wearing sunscreen, broad-brimmed hats, and eye protection against cataracts. We certainly hope so.

POSTSCRIPT

Contrary to conventional lay and medical belief, farm children with early exposure to plant and animal allergies do *not* appear to be at any disadvantage. They may actually benefit from early childhood exposure to barn and field allergens from plants and livestock, in contrast to children raised in air-conditioned, clean, urban environments. Pediatric research indicates that their immune systems build up tolerance for a great variety of rural antigens which paradoxically protect them from developing childhood asthma.

This is another example of scientists trying to sort out *alleged* risks of farming from *true* risks—infections from noninfections, susceptible from resistant segments of the population, acute stress from chronic stress, reproductive fears from reproductive risk factors, and preventable from not-so-preventable hazards in agriculture.

Chapter 10

Raspberries and Tomatoes: Golfers and Growers

SCENE ONE

Golfers' Luncheon

"Doc, I'm really worn out from this diarrhea. How can it last for ten days with all the antibiotics I've taken?"

"Well, Rosemary, at least you're getting better. You know you're the sixth case I've seen in this office. I'm waiting for the state lab's diagnosis, any day now. They must be doing a thorough battery of tests to take this long. You think that you and your friends got the bug at a country club luncheon?"

"Well, doc, I've made a list of everyone I remember at the four tables at the luncheon. It was easy because we had played nine holes in foursomes, a total of sixteen golfers. Our foursome came in second; we got the red ribbon."

"Get to the point, please. I have other patients waiting."

"Well, you see, we all had the same chicken salad with a choice of soup or mushroom."

"Okay, okay, so . . ."

Rainbows in Washtubs

doi:10.1300/5851_10

"Well, Doctor, eight of us developed this awful cramping diarrhea, with low grade fever and, you know, blood and mucus in our stools."

"I really don't have much time for this," Dr. Palmer interrupted again.

"Wait, doctor, please hear me out. When dessert time came, the chef apologized, saying that he ran out of fresh raspberries to go with the pound cake and lemon sherbet."

"Okay, okay, already," said Dr. Palmer impatiently, as he got up to go to the next examining room.

"Wait a little minute," Rosemary said as she touched his sleeve. "The other tables had to eat blueberries."

"So." The doctor tried not to look as exasperated as he felt.

"Don't you see, Doctor? *None* of the golfers who ate blueberries got the diarrhea!"

If only Rosemary had gone to medical school, she would have made a superb epidemiologist. Sadly, despite his clinical acumen, Dr. Palmer would not.

SCENE TWO

Salmonellosis from Farmer Isaiah's Famous Tomatoes

"Who called this emergency meeting?" someone in the group of hastily assembled Hope Island farmers asked Isaiah. The men were all tanned, weather-beaten, veteran tomato growing neighbors. Sipping beer and iced tea in the evening, a dozen of these old time friends had been summoned to meet at Isaiah's packing shed on short notice.

"I did, Bozo!" asserted Isaiah, pulling the baseball cap down over his cousin's forehead.

"Watch who you call Bozo, Junior!" he snapped back good-naturedly.

"Listen, guys, let's get serious!" Isaiah called for some order. "We're all in this mess together. I'm on the hook now. These Atlanta characters are coming here from CDC. Your turn may be next! They want to confiscate my entire tomato crop. That will kill the year for me. Something about salmonella in my tomatoes turning up in Minnesota and Illinois! I think someone in Atlanta or Washington is out to get me, or punish all of us here on the Island."

"How in the world," asked his cousin, "can they accuse you of food poisoning in the Midwest?"

"That's really weird," Isaiah agreed. "Maybe we need to contact our association lobbyist in Washington. Maybe we need to call our lawyer and sue the bums!"

The farmers were all talking now and still trying to get Isaiah's story of woe straight.

"Last week, a federal agent from Atlanta calling himself Doctor came to my farm, in a car with a government sticker, bold as brass, saying, 'I'm here on official business from the U.S. Public Health Service.'

"He sticks a card in my face and orders me to show him my entire packing operation, from A to Z. We were just finishing the season's crop. Not sure of what to do, without getting the sheriff out here to read me 'the law' on food safety, I obeyed the fella. Next thing I know, he's sniffing around the shed, checking out the worker's toilet, sink, soap dispenser and towels and the conveyor belt leading to the wash tanks. He even looked for bird droppings. Believe me, my loyal, regular packers were nervous with him around, especially when he opened the trunk of his car and took out some chemical testing outfit. With hardly a word of explanation, nary a 'please' or 'thank you,' he starts taking water samples in test tubes from all three tomato wash tanks. Acting real mysterious, running some color-dye-tests, and saying 'so. . . .' Real mysterious and huffy, you know. By now, all of my workers got nervous and left as quick as they could, leaving just the two of us."

"Get to the point, Isaiah, for goodness sake," demanded another of the growers.

"My blood pressure went up and up with me and him alone in the shed. I was so angry, fit to be tied. Estelle came in a hurry from her kitchen and told me to calm down or I'd have a stroke! I wanted my shotgun to drive the critter off. Estelle tried to reason with him and told me to call Dr. Jim Robins (the Clemson vegetable expert) right away. I had his home number and rang him, late as it was, almost midnight. But I couldn't wait till morning. Good old Robins says, 'I'll be right over, Isaiah, hold him there until I can get there. I'll try to figure out what's going on.' I said, 'okay, just as you say. I'll tell him you're our expert and you're coming here pronto. But to tell the truth, if he gives me any more federal gaff, I might just finish my grandad's battle with the Yankees, with some birdshot where it might do some good!'"

Isaiah's story did not lack local color, and he certainly held his neighbors' full attention now.

Here it is, only a week after the CDC inspector episode at Isaiah's famous tomato farm. All the Low Country vegetable farmers (tomato, asparagus, lettuce, soybeans, peppers, and squash) are now gathered in the annex of a local seafood restaurant. The air is abuzz with animated and tense conversations. The Washington, DC, lobbyist arrived by jet from the capital to assure the Tomato Growers' Association that he had all his connections working on their behalf.

When the shrimp and oyster dishes are cleared, the lobbyist makes a few more remarks and turns the meeting over to Dr. Jim Robins and me for our comments. Dr. Jim, the postharvest research and extension specialist, had insisted that I accompany him to answer any salmonella illness questions as a university physician and infectious disease epidemiologist in the agromedicine program.

The local and regional health department officers are notably absent. Years of ill will over pesticide regulation, water quality, land conservation, and wildlife preservation issues, have driven a wedge between the growers and the public health establishment that will require a generation or two to change. After a few introductory remarks by farmer Isaiah, Dr. Jim takes over. He explains that the CDC inspector is in fact a qualified doctor, a very competent DVM and a safety expert from Atlanta. Jim is greeted by a chorus of hisses and grumbling.

"Yes, he is qualified," Jim insists, "and he's actually trying to do his job, protect the food supply for the public. Isn't that what we all want? Quality food that's safe to eat."

A murmur of agreement ran through the group. "As I said, he has all the doctor papers and credentials as an expert in food safety, but," a long dramatic pause followed. "But, actually, he wasn't qualified to do the correct chlorine test on Isaiah's tanks. I had to show him how to measure the residual chlorine correctly . . . with his chemical kit and with mine. . . ."

This was received with a round of applause, whistles, and shouts of "I told you so's" and a few "Atta-boy Jim's."

"But hear me out," Jim continued. "Isaiah's residual chlorine level just barely passed the minimum recommended standard. And the chlorine rinse is critical, as you all know. So far, it appears that no official federal penalty can be filed." There was a stirring of signs of relief.

"Nevertheless, Dr. Stan, our friend from the medical center, here has something very, very important to share with all of us."

Slowly, I rose to address the group of angry, frightened, and puzzled farmers in a very unfamiliar, emotion-charged atmosphere. I tried to speak calmly and steadily to overcome their distrust of the microbiologic information I was about to deliver—a technologic break-through that was just beginning to revolutionize diagnostic bacteriology in the United States and overseas.

"Gentlemen, you know of the advances being made in your field of vegetable crops with the help of lab science. Well, I'm here to tell you that only very recently has the latest technology at the CDC been able to use DNA-type fingerprinting of any bacterial strain that gets from a particular grower's farm to the consumer's table and to the patient in the hospital. And, believe it or not, it's only one small step from the patient to a lawsuit for damages."

Isaiah moaned as the microbiologic news began to sink in.

Then all the farmers gathered around Dr. Jim to ask him to check their wash-tank equipment for chlorinating rinse water at a tomato-safe level. "Whatever it costs; as soon as possible," they all chimed in, eager to take advantage of the free extension services of Jim, their trusted Clemson scientist.

SUMMARY

Food Safety Issues Never Sleep and Begin to Make the Front Pages

The golf luncheon and the tomato growers' meeting illustrate some of the more interesting incidents involving our state-funded, inter-university agromedicine program, linking experts in agriculture with resources at the university medical center to diagnose, treat, and hopefully prevent at the earliest possible time in our state of South Carolina. When the foodborne outbreak begins, local, state, and regional health departments, the CDC, and responsive food industry scientists are all increasingly involved in limiting morbidity and mortality.

The first scene, the mysterious women's golf luncheon, actually confirmed the theory of the talkative, but indomitable patient, Rosemary. She actually did the essential epidemiologic detective work by getting histories of the meal from sick and non-sick attendees, while their recall was fresh. Rosemary, without a computer, or attack rates, or food analyses, guessed that the critical exposure of the victims was eating *raspberries,* not blueberries. Weeks of stool testing from Columbia, SC, and the reference lab in Atlanta confirmed the presence of cyclospora as the antibiotic-resistant cause of the dysentery. The trace-back epidemiology of suspect raspberries in this and a few other similar outbreaks led the CDC to investigate a chain of restaurants, retailers, wholesalers, and importers from the fields of Guatemala. At the source farms there, cyclospora samples were directly linked to poor sanitation and sewage conditions adjacent to the raspberry fields and the workers' poor hygiene (lack of outdoor toilet facilities). The unique properties of the raspberry itself added to the futility of the postharvest rinses that the crop ordinarily received.

Each raspberry has dozens of tiny crevices for the parasite to cling to and to avoid the cleansing operation. Cyclospora are uniquely capable of surviving refrigeration and hours of transit from farm to table, as proven by the widespread, multi-city outbreaks of diarrhea.

The tomato-salmonella dilemma in the second scene was not the first to hit Isaiah's famous Low Country farm. Unfortunately, he had more than one shipment of his otherwise prime tomatoes reach their destination markets in deteriorating condition, unmarketable. As a result, his losses of customers and income were threatening his survival as a grower. Somehow, over the years, Isaiah had cut a few corners in maintaining his packing shed tomato rinse-tanks; his temperature and pH regulation were inadequate to maintain proper chlorine levels during the long and busy hours of the harvest. Dr. Jim, after the fact, was able to correct his tanks by installing newer pH and temperature monitors to prevent a recurrence of the salmonella problem.

My limited role in the tomato farm incident was to back up Jim in defusing the anger and resentment of the farmers against federal bureaucrats. How? Simply by sharing the truth with them. I explained that with the latest tools of DNA and PCR probes for tracking specific Salmonella strains isolated from patients' stool by health departments in the Midwest, with samples of tomatoes from wholesale shippers, and with trace-back investigations (farm-to-table), Isaiah's farm in South Carolina was pinned down as the source of two years of different species of salmonella in tomato episodes. Thus the CDC's timely investigation of his farm was crucial and justified. (By the way, there are over 800 to 1,000 different species of salmonella lurking through the food chain and enjoying our planet's abundant animal hosts.)

Molecular-fingerprinting of human and tomato samples by the CDC had to be explained clearly and forcibly to the incredulous, tough-minded, growers whose livelihood was being threatened. This incident occurred several years before DNA testing became so popular on TV crime shows and thus, part of everyone's vocabulary.

Happy Ending: Believe It or Not!

Far from being unwilling to learn from the salmonella incident, the growers and their growers' association followed Dr. Jim's advice. They raised research funds to support analytic studies at a land-grant research facility, measuring the capability of fresh-picked tomatoes to harbor and retain microbial growth, even after routine rinsing and refrigeration.

This work resulted in an important paper in a national agricultural journal and a collaborative report by Dr. Jim, together with the CDC's expert investigation, in the *Journal of Infectious Disease.*

All's well that ends well.

I was pleased to be of some assistance (medically speaking) to Dr. Jim, to Clemson, to MUSC, and most of all, to the dedicated growers, their families, and their hard-working pickers and packers, whose livelihood depends on safe, quality produce being shipped and consumed at markets and restaurants nationwide.

Food safety issues never sleep, it seems. Even health-promoting foods, such as bean sprouts were able to carry dangerous E. coli 0157:H7 in the school lunches of thousands of Japanese children in one outbreak, resulting in severe kidney infections (Hemolytic Uremic Syndrome). Up to 10 percent of children under age ten years can become hospitalized with the food-borne bacterium, and too many victims do not survive.

Chapter 11

Diet, Cancer, and Birth Defects: Nutritional Mysteries

Can diet cause cancer?

Can diet prevent cancer?

The answer to both questions is a resounding *yes.*

Ten to twenty years ago, the answer would be more ambivalent, but now, thanks to the combination of laboratory studies, epidemiologic surveys, and intervention trials, it's no contest. The key role of naturally occurring carcinogens and anticarcinogens in the diet outweigh, by far, early concerns regarding artificial sweeteners, coloring agents, and tiny amounts of artificial preservatives.

We are all currently the beneficiaries of the teamwork between nutritionists, the food industry, epidemiologists, and public health educators trying to restore a healthy balance between exercise and diet to lower the risks of some major cancers, including cancers of the GI tract (colon, stomach, esophagus, rectum, liver, multiple polyps, etc.) and the endocrine reproductive system (breast, prostate, uterine). Lifestyle and behavioral modification approaches to cancer prevention target obesity and alcohol and tobacco abuse. The hereditary component and aging-lowered immunity factors carry their share of our cancer-risk profile, as well.

Rainbows in Washtubs

doi:10.1300/5851_11

SCENE ONE

During World War II, under Nazi occupation, the entire civilian populace of Holland and Denmark was subject to near-starvation. Clinical and epidemiologic observers expected a rise in classical malnutrition illnesses, including tuberculosis. Some vitamin deficiencies did become prevalent. What about the so-called chronic diseases, including arteriosclerosis, coronaries, strokes, diabetes, peptic ulcer, breast cancer, and, of course, obesity? To the surprise of Danish and Dutch observers, a paradoxical phenomenon happened: these chronic diseases declined year after year as the occupation wore on. Shortage of tobacco and alcohol for oppressed civilians added to the shortage of modern civilization's "blessings." Epidemiologists used the European system of comprehensive and extremely detailed vital records to document some of these contradictory World War II findings: fewer cases of chronic disease from near-starvation.

Decades later, nutritionists and doctors warn of the ominous "Four Horsemen" of aging in the United States—obesity, diabetes, hypertension, and arteriosclerosis—as consequences of overnutrition, under-exercise, and poor luck with one's genes.

SCENE TWO

Dr. Ernst Wynder, later a senior epidemiologist at the American Health Foundation based in New York, began his career as a medical student at Washington University School of Medicine, St. Louis, Missouri, when he got a summer student's job in the laboratory of the distinguished chief of surgery Dr. Evarts Graham. Against all odds, Dr. Graham had succeeded in removing the cancer-ridden left lung of a middle-aged life-long smoker. This feat led to the acceleration of thoracic surgery across the country and overseas. At the same time (1950-1951), Dr. Graham was troubled by the history of many of his less fortunate patients, who were not candidates for surgery (late lung cancer with metastases). Case after case of lung cancer could not be linked to occupational exposures (radiation, coal dust, or radioactive ore) but showed patient cigarette use at one and a half to two packs a day. Users of chew tobacco and cigars wound up in oral surgery.

Ernst was a very bright, hard worker, who reportedly needed little sleep and allowed few distractions from his studies. He responded enthusiastically to a systematic review of Dr. Graham's charts of lung cancer patients and noncancer patients with a history of smoking and occupational exposures. Across the Atlantic, two British doctors (Richard Doll and Bradford Hill) were exploring the same smoking/lung cancer hypothesis—cross-Atlantic serendipity.

SCENE THREE

Turn the calendar forward twenty years. The scene shifts to Lin Xian, China where a small team of American researchers from the National Cancer Institute in Bethesda, MD, are visiting the wards, villages, and homes of peasants suffering from the highest rates of esophageal cancer ever documented anywhere in recent times. Curiously, the local veterinarians and poultry scientists could also point to rates of gullet tumors in chickens just as high as the human rates of gastrointestinal cancer. All of this suggested some hidden environmental or dietary carcinogen causation in this remote agricultural province, located far from the industrialized cities in China with unregulated pollution of air, water, and soil.

A rural cancer waiting for investigation? More importantly, a province waiting for some kind of effective prevention. This tragic situation in China called leading epidemiologists, nutritionists, and oncologists to visit and collaborate with local and regional medical schools and health departments in the search for the most likely cause.

As of 2007, almost thirty years of clinical, environmental, and nutritional investigation have yielded important clues to causation. Dietary deficiencies of fruits and vegetables in the diet, selenium deficiency in the native soil and water supplies, mycotoxins in corn (aflatoxin) and pickled vegetables have been considered as causative or triggering agents. Genetic predisposition (HLA typing) of the cancer patients did not differ significantly from non-cancer residents. Other genetic cancer markers remain to be discovered.

The massive scale of the population problem led to the world's largest nutritional intervention study, begun in the 1970s and continuing to date. Nearly 150,000 total participants in nine different interventions using randomized, double-blind, placebo-controlled designs have tested whether vitamins and/or mineral supplements, given singly or in combination, could prevent total or site-specific cancer. NCI researchers and Chinese health workers have teamed up for this historic nutritional program.

Beta-carotene, alpha-tocopherol, selenium, and retinol have been administered as micro-nutrients to supplement the diet in Lin Xian. Not surprisingly, this geographically localized cancer epidemic aroused a wave of renewed interest in dietary approaches to cancer prevention in the United States, Europe, and Asia.

Epidemiologists keep looking for such opportunities in high-risk populations to shed light on human diseases. In Kenya, a surgeon with a surprising knack for epidemiologic observation found low rates of colon cancer in natives who enjoy a high fiber, low-fat diet. In southeast China a rare form of nasal cancer occurs in children who were served traditional pickled fish as part of their diet.

Geographic and demographic patterns of chronic disease (cancers, arteriosclerosis, metabolic disorders including thyroid goiter, and diabetes) intrigue the inquiring minds of international epidemiologists. The search for genetic, environmental, nutritional and infectious agents continues.

SCENE FOUR

"Stan, would you be interested in studying cancer of the esophagus in coastal South Carolina?" The phone call from Dr. Storey at NCI came to me.

Knowing how our department chairman would welcome any prestigious federal cancer grant, I replied, "I guess so, Bob. I assume you want a case-control nutritional study of patients coming to clinical centers for diagnosis and treatment."

"Of course, Stan, it will take a couple of years to accumulate enough cases. We want to test some of our hypotheses about causation coming out of Lin Xian, China. Your patients in South Carolina are mostly poor rural African Americans, I believe."

"That's right."

"The role of diet in rural poor Chinese and African Americans would help prove our crucial deficiency theory—not enough antioxidants—operating across two different racial populations," Storey continued with considerable enthusiasm.

"You folks at NCI know that older studies here in the coastal region pointed to possible carcinogens in bootleg whiskey. Poor folks made it from moldy corn mash in homemade stills for years, even generations. Some even got lead poisoning from the stills."

"Stan, the hardest and saddest part of your job will be to find the cases as soon as possible. They usually wait to seek help when they can hardly swallow. Their cure rate is terribly low. Your team will have to interview poor, sick, illiterate, and dying men before they are too weak to give answers to our detailed fifteen page history—our standard questionnaire. We are equally interested in lifelong histories of smoking, using snuff, chewing tobacco, and rinsing with mouthwash, as well as all forms of commercial or homemade alcoholic drinks." Dr. Storey was making certain that the challenge of interviewing dying men would not be underestimated by me and my team.

In Charleston, at our medical center, I was more fortunate than I deserved. I recruited Miss Tate, a recently certified nutritionist, who had grown up in rural Georgia. She knew more about the social culture of young and old African Americans, their diet, their recreational drinking, and their tobacco use than anyone we could have recruited from NCI in Washington, DC. Most of all, she had compassion, and even the most tired and weakened patients responded to her down-home accent and kindness of heart.

As the months of study rolled by, Dr. Storey's site visit to our center confirmed our impression that Miss Tate's interviews were accumulating very high quality data for case-control multifactorial analysis. Her empathy and patience with each dying patient worked beyond our most hopeful expectations.

Results of our study pointed to the following causal factors:

- Deficiencies of fresh fruits and vegetables in their diet
- Abuse of alcohol and tobacco from early teen years
- Binge starvation on weekends, when their drinking was most intense
- Poverty-driven use of homemade brews, probably laced with a variety of natural mycotoxins from cheap corn mash, plus toxic lead from homemade stills

From Lin Xian, China to Gullah country in the southeastern United States, the protective role of diet in preventing gastrointestinal cancer was being confirmed in human, case-control studies among different ethnic groups and different soil conditions. That's international epidemiology at work.

SCENE FIVE

A five-star hotel in New York City is the site of a three-day research conference on "Preventable Causes of Human Cancer and Therapeutic Advances." I am in the audience of clinicians, epidemiologists, and public health workers, awaiting the plenary address by Sir Richard Doll.

Dr. Doll had pioneered cancer research in the United Kingdom by focusing first on occupational exposures (woodworker's nasal cancer) and radiation exposures (miners and radiation therapy patients), and later on smoking and alcohol abuse (lung cancer, liver cancer, colon cancer), leukemia and nuclear plants, and so on. For his incredible discoveries, he was knighted by the Queen and honored with the most prestigious international prizes awarded for pioneering cancer epidemiology.

Immediately, Dr. Doll charmed the audience with a gentle but witty account of an encounter with a Manhattan taxi driver who had trouble with his accent. Then, with a few select slides, he summarized the findings of his team's three-year study (meta-analysis) of the world's published literature on case-control, geographic, and occupational studies of all the major causes of cancer deaths. The results were brief, clear, and compelling. Human cancer causes fall into two major categories: internal (two-thirds of cases) and external (one-third of cases). Internal causes include the following:

Genetic risks (oncogenes)
Aging (inevitable decline of immune defenses and cellular repair mechanisms)

External causes include the following:

Dietary deficiencies
Dietary excess, manifesting as obesity

Alcohol abuse
Tobacco abuse
Lack of regular exercise
Occupational and environmental carcinogens

Doll explained that his comprehensive research on environmental and work exposures (with the exception of UV light) show a *minimal* component of total human cancer incidence. Doll cautioned that so far it is difficult to find cancer evidence to justify our growing concern over industrial air, water, and soil pollution of our planet, with the possible exception of the damaged ozone layer and UV-triggered skin cancers.

In summary, Doll stated that two-thirds of the external causes of human cancer could be mitigated effectively by a balanced *diet* of lower calories with abundant fruits and vegetables, regular *exercise* at all ages and avoiding alcohol and tobacco abuse. That leaves the majority of human concerns (two-thirds of the total types of cancer) waiting for research breakthroughs in aging, immunity, cellular repair, and the human genome.

Primary Prevention: Lifestyle choices will work.

Secondary Prevention: Early detection, early treatment, and follow-up will work!

After his tour-de-force presentation, the entire audience of scientists rose to applaud this remarkable epidemiologist with his life-long career of innovative research. Colleagues, former students, and admirers lined up to greet him at the coffee break, while media reporters left the room to meet their deadlines.

SCENE SIX

It's Saturday morning in the ballroom of the five-star hotel. I am the "caboose," the last of the last presenters on the final half day of the cancer-epidemiology meeting.

The nearly empty auditorium indicates that most of the scientists are checking out of the hotel to catch early planes out of LaGuardia. I clutch my few slides for my brief presentation of a case from South Carolina. I can't help being nervous, even though there are only a dozen or so people in the audience. They're probably friends and relatives of the preceding five speakers. I am speaker number six. As each speaker before me finishes, gathering his slides, another tiny portion of the audience vanishes.

I am wondering why I bothered to submit my abstract "Osteogenic Sarcoma with Craniofacial Anomaly in Two Unrelated Girls" (with maternal history of breast cancer; a new manifestation of the Li-Fraumeni syndrome of birth defects predictive of later cancer). My slides were simple facial photos, craniofacial and extremity X-rays, plus family histories of the two remarkable youngsters with bone cancer. A genetic expert in Geneva would soon include the clinical syndrome as the "Schuman-Burton Syndrome" in his encyclopedia of genetic malformations of medical importance from the world's literature.

As the house lights were turned up, in the hollowness of the regal hall, the host of the session asked politely if there were any questions, from the few persons remaining in the audience, before the cancer proceedings officially ended. A hand went up from a tall, white-haired gentleman who stood alone with a broad smile, and spoke with a British accent.

"I want to thank Dr. Schuman for a fine contribution to our understanding of genetics predictive of later cancer. I believe molecular chemists will, in the near future, come up with a number of specific predisposing 'lethal genes' to explain many of these unfortunate occurrences in cancer families." Dr. Doll gathered his briefcase and waved to me, isolated on the podium, as he also departed for the airport limousine.

Soon, laboratory scientists would be linking affected cancer/birth defect families with the p53 part of the human genome, just as the great Sir Richard Doll predicted. His kindness to a lonely investigator on a late Saturday morning in Manhattan would not be forgotten, as I was added to his legion of admirers around the globe.

Footnote

The causual gene anomaly was discovered, as Sir Richard Doll predicted, explaining the Li-Fraumeni syndrome, which included the Schuman-Burton variant of the familial cancer/birth defect disorder, and many other variants of cancer kindreds.

SCENE SEVEN

Here I am, in Atlanta at the busy headquarters of CDC/Epidemiology Branch, waiting for Dr. Urban to get off the phone.

"Stan," he says cheerfully, "what can I do for you?"

"Well, as a fellow pediatrician, I have to ask how you 'drifted' into epidemiology and into the CDC."

"Same as you, probably, Stan," he smiled. "I just got bored and restless doing routine ped clinics and hospital rounds, although I loved working with kids. I felt like my brain was half asleep, treating the same old cases of otitis, diaper rash, eczema, cow's milk allergy, and bedwetting. You know what I mean?"

"Well, you're not bored now, are you?" I asked, rhetorically. "Tell me what I need to know about birth defects. I keep getting calls from distressed farm families in South Carolina telling me that using agrichemicals caused their child's birth defect—cleft lip or club foot or hypospadias or one blue eye and one brown."

"Stan," he replied, "I've been doing birth defects and running the greater Atlanta registry for years now. And I've been comparing notes with my colleagues in Europe. They have some fine data in their birth defect registries to add to ours at the CDC. The truth is, Stan, that two-thirds of human birth defects are still totally unexplained."

"Even with chromosome analysis?" I asked, "And metabolic study, and genetic screening?"

"Yep," Urban added. "But the one-third of birth defects we can diagnose still keeps us busy."

"Are pesticides part of that one-third?" I ask.

"No, Stan, as far as we know—despite worldwide use and exposure to DDT and herbicides like 2,4,5-T, and the organophosphates, and ground water contaminants—there are no signs of fetal damage comparable to the Big Six."

"What are they?" I ask.

"Radiation, rubella virus, heavy metals, vitamin A poisoning, cancer chemotherapy of pregnant moms, cytomegalovirus virus, and that's about it...except for the Big Seventh: it's now dietary, Stan. We're really excited about the ability of folic acid supplements to prevent recurrent neural tube defects.

"The British got on to it way before us in the U.S. We've got to catch up with them. It has to do with a lack of green and yellow veggies in the diet in very early stages of pregnancy and, in fact, in the weeks before conception. It's a critical component in iron metabolism, and if that goes awry, the result is devastating failure of embryonic development of the spinal cord and the brain of the fetus."

"Do you mean that vitamin B complex with folic acid can actually prevent high-risk moms from having a second baby with birth defects?"

"Absolutely," he assured me. "In fact, early results from the U.K., Hungary, and Czechoslovakia are impressive. In Europe, they require folic acid to be added to all flour."

What a simple way, I thought, *to reach all the starch-consuming girls and women, before they can even think of becoming pregnant!*

"Stan, the results of folate supplementation are truly astounding! Neural tube defects are down by a third!"

"When will the U.S. get on the bandwagon?" I asked.

"Soon, I pray," he answered. Then I knew Dr. Urban's service-time at the CDC more than made up for his absence from pediatric emergency rooms and hospital rounds. Here was another pioneer of nutrition, another Joseph Goldberger, making the U.S. Public Health Service and the CDC the envy of other countries around the globe.

Now, I could answer some of the distressed farm families and their obstetricians and pediatricians, explaining that two-thirds of babies' birth defects are still unaccounted for. Nonetheless, improved nutrition, a balanced diet with plenty of fruits and vegetables, should be routine for all expectant mothers and their daughters.

In terms of reproductive health, chemophobia, defined as irrational fear of chemicals (such as environmental exposure to low dose pesticide residues, only measurable in parts per billion and easily analyzed by chemists in our soil, air, or groundwater), is not helpful. Such environmental concerns are understandable but must be based on good science. The search for specific toxic or chemical teratogens needs to go on, especially with the new DNA discoveries of the human genome. The absence of environmental pollutants as teratogens should relieve some of the guilt parents feel when a farm child is born with a major birth defect.

In fact, contrary to popular media fears, 3 percent of all human births have *minor* birth defects (a strawberry mark, etc.), and about 1 percent are born with *major* defects (heart defect, etc.). That means that about 95 percent of human term births result in perfect newborns—a statistic that has not wavered since birth records have been kept (since before the Industrial Revolution and sustained throughout this modern age of urbanization and better medical records).

Chapter 12

Hog Farm Miasma: Case Studies

SCENE ONE

My office phone rang, first thing in the morning. "Dr. Stan, this is Tommy at Ridgeville County. I have a medical problem for you guys at the medical center. My phone is ringing off the hook here at Clemson Extension. Guess what? One is Farmer Mike, one is Homemaker Sue, one is Mike's lawyer, and one is Sue's lawyer. Four calls within the hour, each about Sue's son, Kevin. It appears that Kevin is in the hospital with a serious case of pneumonia, which is more serious than the doctors thought. Some kind of abscess, unusual for a healthy teenager, they say. Sue is convinced that Kevin is being poisoned by the air drift from Mike's corn and hog farm, especially the foul odors from his brand new hog pond. While the bills are piling up from the hospital, she's hoppin' mad and got an attorney taking her case to sue Mike for medical costs, plus! Mike's attorney is working on the defense. Have y'all heard of hog farm air pollution causing pneumonia?"

Rainbows in Washtubs

doi:10.1300/5851_12

"Not really," I answered Tommy. "But my staff and I will look into it with our pulmonary medicine specialists, okay? And we'll get back to you."

"Thanks a lot, Doc. You know this is a heck of a time for me to have to deal with this. We have a county fair and farm city week to organize. Dang these lawyers! Ridgeville used to be such a neighborly place. Now it seems everybody has to have a lawyer to sue somebody!"

"By the way," I tried to make it clear to him, "I want you to know that I, personally, do not testify as an expert witness on any agromedicine case, unless I am subpoenaed. The lawyers can always hire some outside expert from a toxic torts list of professionals to help make their case."

SCENE TWO

While I canvassed the literature on hog farmers' pulmonary diseases, our seasoned chemist, Archy, contacted Tommy's staff to get an idea of which agricultural chemicals Mike's operator would be using at this time of the year. Given the most likely toxic chemical list, our lab could get ready to test swab samples of the interior of Sue's house for chemical evidence of air drift from the farm. Soon, we knew, we would be asked to make a home visit and the lab would already be prepared to run samples from the interior baseboards, carpets, drapes, kitchen counters, and even vacuumed samples of dust from the teenager's bedroom and study.

SCENE THREE

"Look at these bills, just look at them!" Sue insisted, as she led our small team (our chemist, an assistant, and me) to an impressive stack of overdue bills from the doctor, pharmacy, hospital, plus many more from her attorney.

The small home for parents and two teenagers (the sick son and healthy daughter) was modest, neat, and simply furnished. The father's work clothes for the gas company were neatly starched, ironed and folded, along with the family's laundry. She thanked us for coming and gave us a chronological history of the onset of her family's symptoms, her son's pneumonia, and his medical visits.

"Everything seems to begin with our neighbor, Mike, opening up a new hog waste disposal system. His pond recycles animal waste to be used in his corn and soy bean fields as slurry-fertilizer. Today you can't smell the foul odor, but believe me, it can knock you over when the breeze comes down the hill to our home. It's not just Kevin's asthma. He's had that for years. But this pneumonia thing is too much. It's the last straw.

"Thank goodness, Sissy has been okay and has not missed school. But Kevin is being held back a grade because of his missed days. Mike and his family used to be such good neighbors. But that was then, and this is now. My lawyer says we have a good case. We sure need a judgment to cover all these bills. We're not made of money."

SCENE FOUR

I tried to appear more knowledgeable than I was, but Archy was making some headway on improving my limited toxicologic gleanings.

Scratching our heads, Archy and I gathered the evidence to share with our pulmonary expert, Dr. Marks, who was coming over that afternoon. He had already contacted the local physician caring for Kevin and reviewed his chest films and cultures.

"As far as I can tell, Stan and Archy, this is a run-of-the-mill pneumonia case, probably staphylococcal, complicated by pleuritis. Nothing unique about it. Nothing suggests chemical damage if that's what they're looking for. I've treated a lot of farmers with acute ammonia and nitrate and methane inhalation in the past. No evidence here, medically speaking, of chemical damage."

After a long pause, Don, our junior assistant, who meticulously labeled all twenty-four samples for Archy at Sue's home, spoke up, "You know, I had more free time with Sue, than you fellows. She offered me coffee and baked cookies. So we chatted while you were busy. She's really a nice lady, reminded me of some of my aunts who live near Ridgeville."

Dr. Marks politely excused himself and started to hurry off to make rounds at the medicine ward.

"Go on, Don," I said. "I didn't get much of a detailed history from her."

"She didn't volunteer much to me either," Archy admitted.

Don continued, "Well, it went like this. Kevin has had mild asthma on and off since early childhood. His younger sister was okay. His dad had it as a child but outgrew it."

"Okay, Don, go on. This history could be important."

"Well, Kevin missed a lot of school but, somehow, made the swim team and got medals for the 200 freestyle. But Sue insisted he go to the doctor every time he caught a cold. One summer he got a job in town at the supermarket and stayed with his cousin in his apartment. All summer he worked and kept up his swimming. He met a girl he liked, bought a used car, began dating, and didn't have to see the doctor at all for asthma checkups."

"Interesting, Don, any more?" I asked.

"Well, maybe I shouldn't be telling Sue's remarks, but all of a sudden, out of the blue, she said, 'You know, poor Kevin, ever since he was knee high, when he couldn't sleep from whatever, he'd climb into our king-size bed with his dad and me. He never outgrew it. We were always a great comfort to Kevin,' she said, 'and he kept his strength up between attacks.' But she said this darn hog farm and pesticide poisons landed him in the hospital."

SCENE FIVE

A year passes, and life moves on, for Sue, for Kevin, for Mike, and for Tommy. Nothing much changes for the folks in Ridgeville.

Kevin graduates from high school, turns down a swimming scholarship at the state university, becomes engaged, and is promoted to assistant manager at the supermarket. His health improves with no more hospitalizations or doctor's visits for asthma. Sue, his mom, gets a job in the high school cafeteria and, using her management skills, is soon promoted to a supervisory role. Nothing comes of the lawsuits from either side.

Tommy, after years at Ridgeville County Extension, is promoted to regional director over three county extension offices. He also earns the Civitan Man of the Year Award for dedicated community service.

Oh, yes, Don, our lab assistant, applies to medical school and is accepted on his first try.

Our agromedicine office begins to receive more inquiries from troubled citizens and extension agents, and even the local media, regarding the "health damage" being done to neighbors who live near the newer, larger, hog farm operations. The legislatures of several southeastern states are drafting bills to limit the size of hog operations and to mandate a "comfort zone" between waste disposal systems and residential homes and water systems. Health surveys (with skimpy data) are conducted by the epidemiologic unit of one health department. One health officer alleges that excess health risks are being suffered by citizens living near the pork growing, air-polluting hog factories. High on the list of symptoms (cited by the very sensitive health department report) are headache, sinusitis, excessive fatigue, nausea, loss of appetite, nervousness, insomnia, and trouble tasting and smelling odors.

In the meantime, a specialist and national leader in olfactory research at the medical school has been studying taste-and-smell side effects of widely used psycho-pharmaceuticals. She is awarded

EPA and NIH grants to extend her work in a national conference of experts on animal waste disposal and possible effects of farming odors on human health. As a senior researcher, she still finds it hard to separate subjective symptoms from objective clinical signs.

At the conference, experts in agricultural engineering and animal husbandry shared their hard-earned wisdom with the medical and public health specialists. The state-of-the-art of large, indoor hog operations and environmentally-sound waste disposal is a work-in-progress. Biologically, nitrogen wastes and odors can be modified and improved environmentally, in the interest of healthier hogs and less onerous wastes. Pilot studies apply bioengineering methods to improve the animal diet, waste-absorbent materials and drainage systems in the barn, ventilation techniques, and neutralizers for methane, ammonia and sulfur dioxide vapors at the source. Altogether, modern hog farming is also a work in progress, it seemed.

Historically, the problem of air, water, and soil pollution from larger and larger farm operations has been accentuated by what demographers call "rural encroachment." Rural areas adjacent to farms are being developed into suburban housing and shopping malls. Consider the fact that ordinary pungent animal and vegetable odors surround farm families for a lifetime without observed health effects. But such agricultural odors are barely tolerated by suburbanites who are more accustomed to motor vehicle fumes. The art and science of olfactory physiology is fascinating in its own right.

The neuro-anatomy and functioning of nasal passages leading directly to neural centers are among the newer frontiers for human research. How does the brain cope with taste, odor, pain, pleasure, memories, dreams, sensual attraction and revulsion, appetite, and conditioning of such sensual information? Consider how alternative medicine practitioners have developed aroma therapy as one of their most popular and effective tools.

Thanks to Ridgeville County and Mike's odorous hog pond, our eyes, no, our nostrils were opened to appreciate a wider world outside our medical center. Communities with rural health issues are evolving along with streamlined hog operations.

The questions are: can efficient farm operations succeed with newer technology on waste disposal, and can suburbanites become more tolerant of their farming neighbors?

Chapter 13

Milkmaid's Shoulder: Women and Children at Risk

SCENE ONE

Strange how we think of farmers in the United States and overseas as predominately men (young, middle aged, or elderly). This image is rooted in our past: agriculture as heavy manual labor. Yet here and abroad, if we really look around, there are increasing numbers of women and children engaged in row crops, orchards, packing, rice fields, and vineyards, working for poor wages, working to sustain a marginal existence on underdeveloped lands.

Well aware of this worldwide neglect of the health of women and children workers globally is the International Labor Organization. Periodically, health alerts are posted to the WHO and the International Agricultural Commission, but little progress seems possible against the competing forces of hunger, economic exploitation, and simple survival in subsistence communities.

In Chapter 3, a very personal investigation led me to observe child labor being exploited in a tomato-asparagus farm in the Nile delta. In this chapter, a few vignettes will suggest the potential adverse health effects for women and children in agriculture. Statistics and demographic trends can be researched through WHO, ILO, and the World Food Bank and their annual reports.

Rainbows in Washtubs

doi:10.1300/5851_13

Vignette One

A family doctor, Dr. Braun, working in a rich dairy region of Denmark, is impressed with increasing complaints of neck and shoulder problems occurring in women of working age. After careful history-taking, the doctor finds that certain larger, modern daily milking operations are reporting more disabilities of this type than he would expect. After all, the newer farms are cleaner, better ventilated, better lighted, and better supervised than the older dairy operations. Finally, Dr. Braun visits one of the implicated farms. Being escorted by the assistant manager, he is shown every courtesy. In fact, each section of modernized machinery and plant improvement is shown to Braun with great pride. When he comes to the rows of milking operations on the highly prized herd, he observes an almost complete absence of men. The women wear hairnets and caps, heavy duty boots, clean waterproof aprons, but no gloves as manual dexterity and speed are at a premium. And these women are good at their jobs, as they joke and chatter with each other over the cows' backs, above the drone of automatic milking machinery.

Braun recognizes one of his patients whom he recently diagnosed with arthritic inflammation of the cervical spine at C5 and C6, a little premature for an otherwise healthy woman in her mid-thirties. A physical exam had revealed exquisite tenderness of several pressure points on her left rotator cuff.

Marianna, the patient, greets him warmly, wiping her hands on her apron, between milking her cows. She slaps her favorite, Missy, on the flank, and tells Braun that she and Missy are good old friends. Trying to get accustomed to the continuous noise of the powerful milking machines, Dr. Braun inquires about her anti-inflammatory medications and cautions her to watch for signs of gastrointestinal side effects.

Marianna smiles and insists that she is sure her arthritis is probably due to the damp, cold early spring season, as her mother-in-law advises her. As the automated milking machine signals that it is ready for the next cow, Marianna sits on her stool to detach the hoses and raises the apparatus above her head to the conveyor belt where it will be sprayed with cleansing solution and returned to the steel rack over her head. A buzzer signals when it is ready for attachment to the next dairy cow.

At this point, with the next cow waiting in her stall, Marianna must reach up with both arms to grasp the sleek, modern milking apparatus. Braun notes that she must grasp with both hands and pull hard to bring it down to the level of the udder, holding it there for attachment.

How many pounds of weight? he wondered. *At least 25 to 30 pounds,* he guessed. *No wonder,* he thought, *that repetitive strain and neuromuscular stress could build up during Marianna's busy work day. Odd,* he mused, *that the milking apparatus was spring attached to elevate overhead, but had to be* manually *lowered with upper body strength to start each milking operation.*

Marianna nodded farewell and returned to her work, murmuring that she hoped warmer, less humid weather would return soon.

Driving home, Dr. Braun wondered how such a modern plant could have underestimated the muscular strength required by an average woman milker to lower the apparatus again and again during a working day. The occasional male dairy worker registered no complaints of neck and shoulder strain, but they usually demonstrated early tinnitus symptoms (ringing in the ears) from avoiding protective ear plugs, provided by management, and instead playing rock music on a headset, "to drown out the noise."

Dr. Braun may have been exposed to sports medicine and repetitive motion disabilities during his college soccer days or perhaps during his family medicine training. But it is unlikely that he had any formal exposure to the art and science of ergonomics that has developed in recent years from the study of heavy and light industrial operations.

"What is ergonomics?" he might ask an occupational medicine specialist, who could reply, "It's improving the dynamics operating between the worker, the task, and his or her tools. It's more than improving computer keyboards for typists and orthopedic seats for long-haul truck drivers, although these are good examples. A well-trained ergonomic team can study, diagnose, and recommend preventive methods for most manual occupations, especially those in repetitive motion jobs."

In advanced and developing countries, there is a severe shortage of trained ergonomic specialists, even though they can reduce the musculoskeletal hazards (acute, delayed, and chronic) of repetitive motion labor. In the Philippines and Southeast Asia, one sees young children and pregnant women, lightly clad, with heavy pesticide-spray packs on their backs treating and over-treating rice fields (without any protective clothing, such as gloves and masks).

Examples of poor to awful ergonomics can be cited in many agricultural occupations in many countries. Milkmaid's knees and milker's nodules of the fingers are well-known effects of wear and tear on susceptible joints. Among the most arduous, painful, and brutal examples is cutting sugar cane or bamboo stalks manually with the machete. The workable lifespan of cane workers is short and disabling, according to the few occupational studies devoted to them.

More encouraging, is an innovative example of ergonomics waiting to be applied. It is based on a University of California study of labor involved in grape harvest at a typical modern farm. A team of experts at the University of California at Davis were invited by vineyard owners to analyze the ways in which harvesting grapes and trimming thick wooden branches could be improved, to increase efficiency and minimize orthopedic problems. Their month-long field studies resulted in several simple, feasible ergonomic improvements, ranging from an improved grip on a hand tool to a chest-high cart for harvesting. Their suggestions were amazingly simple and quite easy to implement for the benefit of the grape workers and the productivity of the vineyard.

Vignette Two

Dr. Robins, a Clemson specialist in postharvest horticulture, is a frequent consultant to far-flung operations in Latin America. He returns to his family of small children appalled at some of the child labor he sees on a variety of fruit and vegetable plantations. "Number one," he asks "how can they work in the fields and in food processing sheds and never go to school? Second, how can managers allow mere boys to stand waving flags at the corners of fields being sprayed with hazardous chemicals from aircrafts? Third, how can migrant worker families in the U.S. allow their preschool children to play in the water of irrigation and drainage ditches contaminated with the run off from latrines and chemical applications?"

"This," Jim explains, "is not high-tech ergonomics. This is simply public health and sanitation for poor children at risk."

On a brighter side, Dr. Vick, a research professor of entomology, has led a number of field studies and field training sessions for peasant farmers to preach the benefits of integrated crop management. His message is how to reduce the need for and the exposure to powerful agricultural chemicals by treating the fields only when a threshold of need is diagnosed, by simple scouting methods. This environmentally friendly kind of farming actually protects beneficial insects and maintains a healthier, consistent crop. In addition, Dr. Vick's team studies biological ways to resist plant invaders. Use of a crop-friendly virus can help control certain nematodes from infesting crops. This type of crop management also limits the risk of chemical exposure for farmers and their field workers.

Vignette Three

The work of several Indian professors sheds light on the problem of poor ergonomics exacting a toll on the daily lives of peasants and poor farmers in Asian countries. They study age-old methods used and honored by the workers, which, with minimal modification, can be improved to lessen musculoskeletal wear and tear. For example, a better handle on a simple scythe, a better cutting angle on a simple plow, even a better-designed harness for oxen.

But times and cultures change. Soon these scientists will need to modify all-terrain vehicles with roll-over bars, as the peasant's trade in their oxen for the latest labor-saving machine.

SUMMARY

In each of the vignettes, teams of physicians and agricultural specialists have demonstrated the ability to see a problem and raise the critical questions (why? or why not?). With dogged determination, they try to find ways to relieve the wear and tear on the worker in ways that are feasible and culturally acceptable in the local setting. Curiosity, observation, caring, and commitment are some of the tools cited in Chapter 2. The art and science of ergonomics (systematic observation, analysis, pilot study, field testing, and applied engineering) are waiting to be applied to the most needy and ancient of mankind's occupations: farming. The savings in efficiency, productivity and worker health are just beginning to be appreciated.

Chapter 14

The Next Generation

SCENE ONE

"Good morning, class," I start the seminar cheerfully. My eyes scan an auditorium full of eager, healthy, shining faces. They are idealistic young men and women casually dressed in colorful college T-shirts, khaki shorts, and comfortable sandals, starting their very first week in medical school. It's Indian summer in muggy Charleston, South Carolina, under threat of the latest hurricane.

"Good morning," some of them answer. The others are noncommittal.

The one-and-a-half-hour session goes smoothly for me, and for my three faculty helpers: Dr. Wells, Dr. Robins, and Dr. Vick. Our theme is Agromedicine: An Introduction. Professor Young is our host and coordinator of the basic science course, Biochemistry 100, which accompanies Anatomy during the grinding first year of med school.

As guest faculty, we are in good hands with Professor Young. He gives us a brief but glowing introduction to the eager but slightly intimidated class.

"Relax," Bill reassures them. "Get ready for some really important information. These gentlemen from Clemson Research and Family Medicine will open your eyes to their kind of work in the

Rainbows in Washtubs

doi:10.1300/5851_14

community—the kind of community you may be serving somewhere in our state in the very near future!

"As I said before," Professor Young continues, "just listen. None of this seminar will be on the first quarter exam."

One hundred fifty notebooks close and retractable pens click off. Students lean back and wait for our first slides. All those pens clicking off make quite an unusual chorus.

I take a deep breath and start my part of the four-part team teaching. I start with my "Finding Rainbows in Washtubs" story of near-tragedy in an underserved cotton-growing county (see Chapter 1). I emphasize the role of the family medicine intern (doing locum tenens) calling me at home on Saturday night from the ER in the rural county.

"Picture yourself in his shoes," I say. "Far from this Medical Center, late at night, giving intravenous fluids to a very sick dehydrated child, holding a small coke bottle, filled with some strong smelling cotton insecticide, with a name you never heard of."

SCENE TWO

Two weeks later, twelve of the freshmen biochemistry students are seated around the conference table adjoining our agromedicine laboratory. They have volunteered to do three months (actually twelve half days spanning three months) of literature research on any topic that fits under the category agromedicine and/or occupational and environmental medicine. Almost half are women, a sign of the emerging role of women in medicine in our school and across the nation.

Each student carries a *monster* book bag, heavy enough for a weight lifter, which they casually swing onto the floor under their seats, making me feel more arthritic than ever. Our round table seminar session of the morning will be repeated a dozen times over a three-month period. There will be plenty of time for discussion as they are introduced to the field of rural medicine and its opportunities for clinical service (curative and preventive). Evidently, the single auditorium introduction by our four-teacher, interdisciplinary team intrigued the freshmen enough to sign up for this elective. This "dirty dozen" chose agromedicine over eleven other competing specialties in research study (ranging from cholesterol metabolism to DNA synthesis to medical ethics).

At our first seminar session, they ask for the criteria required in biochemistry to earn a passing grade. The research required is an *in depth* literature review of any rural topic of their choice. The latest information must be summarized in a two-page abstract format with PowerPoint slides to be presented to their classmates in the seminar room. This exercise will help prepare them for their medical careers, when they are expected to present cases at grand rounds and at scientific meetings.

On opening day, we introduce ourselves informally. I'm an ex-pediatrician, specialist in epidemiology, working in a family medicine center. Dr. Wells is boarded in family medicine and geriatrics and a colonel in the Army Reserve. Dr. Robins volunteers to teach a few sessions relating to his duties in postharvest technology at the nearby Vegetable Research Station of Clemson University. Prof. Vick is just back from consulting for Clemson and the Rockefeller Foundation in the rice-producing provinces of Thailand and Laos. Altogether, we are an unlikely crew to appeal to a basic biochemistry med school class, but we enjoy the challenge.

"Why did you sign up for this?" we ask the "dirty dozen" around the seminar table. More than half indicate some connection to a rural farming experience—an uncle, a grandfather, a summer job for Clemson as a pre-med student. A few are drawn to the possibility of overseas medical ministry. All seem interested in issues of environmental toxicology, wildlife preservation, and environmental protection.

An older student, choosing organophosphates as his research topic, was personally expressing concern about several summers' exposure on his uncle's soybean farm while driving a ground-rig, spraying chemicals. He recalls being soaked through his clothing, at times, and having to wait hours before he could shower and change out of his work clothes. Another student is an avid fisherman. He wants to explore the chemistry and metabolism of mercury residues in the sport fish that he eats regularly.

One of the women was trained in epidemiology as a public health student before entering medical school. She wants to make a career of preventive medicine to help women and children in developing countries in Africa.

Another woman is a surgical nurse, older than the others, who entered medical school to learn more about preventing ovarian cancers. Like the faculty, they are a motley crew, but as the hours and slides go by, we cross some of the divide, the gap, in our ages, our stages of life, our experiences, and our motivations. The mysterious thing called the teacher-student relationship is happening.

SCENE THREE

The years roll by. One class of students is followed by another, and another, until you wake up one day retired from teaching, writing or editing a book or two about the major developments in your small, but important, field of medical endeavor. Retired, you have time to reflect on individual students that you are learning about, who are doing good work in their medical careers in urban and rural areas, here in the United States and abroad.

Dr. Zimmerman is an eye specialist treating children at a medical center and teaching residents the latest technology in ophthalmology. In his spare time, he finds time to serve on medical missionary teams to needy villages in remote parts of India and Africa. There he trains rural doctors and nurses in cataract surgery and simple lens replacement and pushes the parents to increase carotene-foods (Vitamin A) in the diet of their children to prevent blindness.

Dr. Lee has built a network of *four* rural family practice clinics in our mid-state region, where he can deliver primary care to families and only refer cases when necessary to the mega-medical centers in the region's big cities. He has no trouble recruiting like-minded young primary care graduates to staff his family-oriented clinics and the "free" clinics that he serves in his limited spare time.

Dr. Curtis is the older nurse-medical student who went on to specialize in radiology. She writes that she has developed a program to reach rural women in Appalachia for earlier detection of breast tumors through mammograms and self-examination. The importance of better diet, less obesity, and regular exercise is part of her prevention outreach.

Dr. Daly is a pediatrician returning to his roots in another rural region where he serves 4-H camps for farm children, emphasizing farm safety. Valuable lessons focus on how to avoid animal injury, power machinery injury, tractor and all-terrain vehicle roll-overs, lightning, burns, suffocation in silos, sunburn, and heatstroke.

Dr. Eliot became an ER doctor treating a whole range of poisonings from overdoses of recreational drugs (mushrooms, street drugs, etc.) to snake-bites, marine wounds, heatstroke and exhaustion, eye injuries, anaphylaxis, fractures, and head injuries. His nursing staff says he is a tireless worker and role model of dedicated service to patients, especially those he is seeing for the first (and only) time in the ER.

SUMMARY

So, this is the next generation, carrying on the ever-changing types of medical diagnosis, treatment, and prevention that we teachers hoped they would. Did we prepare them adequately as eager fledglings in our classes and on our wards? Perhaps. Will they excel us in their fields of endeavor? You bet they will!

Strange how our lives, our generations, are intertwined. Strange how we are drawn to the land—its bounty, its beauty, its secrets. Strange how we are drawn to nature—its blessings and its hazards. Strange how we teachers suffer from doubts and fears for the future, the student's future. . . . What will happen beyond our brief working careers in medicine?

How will they manage the new technologies or the new labyrinth of medical ethics, economics, and ever-changing patterns of disease? How will they manage emerging infectious diseases, the next AIDS epidemic, and crucial opportunities for prevention?

How will they improve on the doctor-patient relationship: improved history-taking, improved observation, wise use of the new tests and surgical technologies, foster prevention, and positive health behaviors in the families under their care?

Will our students preserve their love of nature, biology, and exploration that attracted them to medicine in the first place?

How will they avoid the boredom and grind of mundane paperwork and bureaucratic stress? How will they react, but not over react, to the occasional over-grateful patient and the over-critical, over-demanding patient?

How can they retain a sense of wonderment and joyful curiosity in the functions of the human organism throughout its growth and development and its potential for longer life and quality of life? Will they appreciate the richness of family, community, learning, and service?

Will they excel us in their chosen careers, using medicine as their window on a wider world?

But, of course!

Chapter 15

Starlings, Elms, and Schoolchildren: Histoplasmosis

SCENE ONE

It's late summer, a mild August day in a very small town in lower Michigan called Milan. Its only commercial remnant of the past is its quaint railroad terminal, a small rural stop on the Wabash line that connects Chicago and Detroit. European immigrants who settled here to farm in the last century saw their cumbersome wagon loads of produce give way to the Iron Horse, pulling cattle cars, lumber, coal, and other cargo to more distant markets. Now the small town only claims an elementary and a secondary school and contains modest homes of families supported by blue-collar jobs at the local federal penitentiary and automobile and refrigerator assembly plants in nearby metropolitan Detroit.

Parking my university car in the shade of old elm trees surrounding the elementary school, I gather my carton of soil samples, in labeled bottles, and my gardening hand-tool, and walk over to the bare earth of the back yard of the school. This is where, during school semesters, buses load and unload rural children for classes. As my professor-mentor and I expected, the soil is bare of any grass, deeply shaded by over-arching elms and time worn by bus tires grinding and pulverizing layers of dusty top soil.

Rainbows in Washtubs

doi:10.1300/5851_15

Dr. Jack Dodge, a seasoned professor of epidemiology and my mentor, has taken me on as a graduate student assistant this summer to work with him. We are engaged in a study of the unusually high histoplasmosis rates in this small community, affecting even the prisoners confined to the federal penitentiary, located on the outskirts of town.

Working together, Dr. Dodge and Dr. Otto Engelke, the county health officer, had focused their concern and attention on the schoolchildren of Milan. Their health might be impaired by growing up in a locale with heavy exposure to *Histoplasma capsulata* spores. This fungus survives as hardy spores in the soil of the Mississippi and Ohio River valleys but is found very infrequently in southeastern Michigan. The health department under Dr. Engelke contacts Dr. Dodge at the university's School of Public Health to study the most threatened and susceptible part of the community, the schoolchildren.

The following questions need to be answered: what is the extent, if any, of their respiratory illness, and what is the source of histoplasmosis in this unlikely village? The citizens and their health department need to know. This is truly an epidemiological investigation, medical detective work, starting from scratch; Dr. Dodge's scientific reputation is on the line, so to speak, and the community is anxious to find some answers. How will their schoolchildren be affected by the strange-sounding threat of histoplasmosis?

If it wasn't for the routine mandatory chest film exams of the federal prisoners taken and read by the astute radiologist during the spring, no one would have suspected that there could be a respiratory-spore problem, causing tuberculosis-like shadows on the X-rays. During follow-up of the federal prisoners and penitentiary employees, specific skin tests were carefully done for the respiratory disease. Most of the inmates and staff were not clinically sick, but the survey revealed mostly positive histoplasmin tests and negative tuberculin tests.

The rate of positive H. tests was as high as 10 to 15 percent. One would expect this among army recruits from the endemic spore containing areas of the Mississippi Valley but definitely *not* in this locale of southeast Michigan, with little, if any, prior soil or human evidence of histoplasmosis.

Dr. Engelke leads an excellent health department team of nurses to skin test all schoolchildren, from kindergarten to eighth grade. The anxious parents give full consent to the school administration. The results of the skin tests with histoplamin are tabulated and charted by Dr. Dodge at the nearby university. The skin test readings reveal a most unusual rising curve of positive tests from the lowest kindergarten rate to the highest eighth grade rate. Fortunately, the only good news was the fact that children's tuberculin rates were very low and unremarkable.

The data clearly suggest an ongoing exposure of schoolchildren to H. spores somewhere in the vicinity of the school, especially since all children who were infected come by bus from different households scattered over the county. But, Dr. Dodge knew from the medical/epidemiologic literature that there had to be a source for the ongoing propagation of infectious airborne spores to reach the lungs of the children and 10 to 15 percent of the penitentiary inhabitants.

Epidemiological field research intrigues Dr. Dodge, who had cut his eye teeth at the army's Walter Reed Institute. He earned high honors for his meticulous observations and analysis of the outbreak of cases of Korean Hemorrhagic fever affecting our troops serving in combat trenches, exposed to the rodent-borne virus, in 1950 and 1951.

When I began my NIH fellowship at the university in May, I was immediately assigned to help the seasoned professor in his investigation of this local outbreak. Fresh out of pediatric solo practice in St. Louis, I had traded my stethoscope and home-visit black bag for a stack of personal household interview forms, designed by Dr. Dodge. I was to help him collect medical and exposure histories from all the households contiguous to the school.

Door-by-door, with clipboards and our friendliest smiles in Michigan's summer heat, Dr. Dodge, a nursing student, and I began to collect over one hundred twenty household interviews. Week in and week out, we rang doorbells. Under a muggy Michigan sun, I had ample time to miss my small, air-conditioned pediatric office in suburban Clayton, Missouri.

The next phase of the Milan study had to focus on an intensive collection of soil samples around the school grounds, and perhaps in adjacent residence yards. Meanwhile, all kinds of rumors were rippling through the community as we strangers from the university canvassed each block with our clipboard of questionnaires to gather any history of respiratory illness or medical diagnosis of TB chest films taken by the residents' physicians.

Later in the summer, we learned that some households were thoroughly confused by another door-to-door interviewer, retracing or anticipating our steps, with his own official-looking clipboard. Our mysterious "competitor" turned out to be an enterprising vacuum cleaner salesman, ripe and ready, with his "guaranteed, inexpensive cure" for histoplasmosis "germs!"

Little did we know how swiftly rumor, fear, and anxiety would spread before us, how the sleeping giant of histoplasmosis (an evil-sounding malady) would grip this community. Local newspaper speculation and gossip had little patience for our time-consuming investigation. Little did we know how property owners were already thinking of selling quickly to escape the strange malady. Real estate values were dropping. The school board was called into emergency session.

When an epidemiologic-infectious disease investigation begins, whether it is illness caused by something at a church supper or an outbreak of a strange rash, there is often little appreciation of the measured pace required to process laboratory and environmental test results. As a well-trained specialist in pediatric practice, I had little appreciation of Dr. Dodge's disciplined approach to the environmental problem posed by the advent of histoplasmosis to a community with no prior documented experience. It is no wonder that irrational rumor has its own way of overtaking science.

"Can we play, too?" Two small cheerful child voices interrupted my soil-collecting endeavor in the extensive bus parking area behind the school. A small boy and his younger sister pulled their red Flyer wagon loaded with plastic toy shovels and pails to my side. Their mother waved to them from her nearby front porch, to let me know that she is watching the stranger gathering soil samples around the school yard in the shade of ancient elm trees.

Ever cautious for possible exposure to spores, Dr. Dodge insisted that I wear a surgical mask while collecting soil samples. With good reason, it seems. The literature on histoplasmosis suggests high-dose concentrations of H. spores accumulate and multiply steadily in very circumscribed environments. *Histoplasma* spores thrive among bat droppings in a residential attic, in a single residential street with soil disturbed by sewer diggers, and in a South Carolina cane-brake field where a bush whacker clearing a small corner lot was hospitalized with fungal pneumonia. With the right combination of soil, climate, and bird or bat droppings, a relatively small area can represent a virtual landmine of respiratory infection; a dose of spores infects in one's lung and spreads to other organs of the body.

The little girl and smaller brother, now my "playmates," casually filled their pails with powdery soil. Seeing my mask, the precocious girl volunteered, "I wore a mask too, last Halloween. This is lots of fun! We play on the swing sets and slides here when school is out."

Driving back to Dr. Dodge's main office with the cartons of tightly sealed bottles of Milan school yard soil, my thoughts began to turn to the beginning of fall classes at the university. Fall leaves were just beginning to turn color and late afternoon chill was settling in. How would I adjust at age thirty-five to graduate school classes? Three academic years of post doctoral studies in epidemiology loomed ahead of me, while my wife would have to manage our slim budget dependent on a fellowship kind of stipend. This summer field study in Milan was already receding in my mind as an interesting exercise, a nice introduction, under Dr. Dodge's savvy guidance, to making every working day in the field count.

Now I anticipated the reality of my uneasy shift from clinical pediatrics in St. Louis practice to being an overripe student again. Now I have to face graduate courses in biostatistics, virology, environmental pollution, and laboratory diagnosis. Did I make a *big* career mistake leaving my small but comfortable private solo practice in suburban Clayton, Missouri?

No wonder I was beginning to feel anxious and less secure than ever, thinking of summer in Milan a mere distraction.

SCENE TWO

Fortunately, classes in the fall turned out to be not as tough as my first year med school anatomy and biochemistry courses. Dr. Dodge proved to be a steady pilot at the academic wheel—a kind but no-nonsense role model and mentor. I had two fellow grad students a few years younger than myself, both MDs, to share our educational post-graduate experience. Both were married and just starting their families. Surprisingly, while we poor graduate students were raising youngsters in a very modest blue-collar neighborhood, brown bagging our lunches to save money, and bicycling to save gas, our wives did not complain. They loved the informal camaraderie and neighborly backyards with swings and slides for the children.

We were told by successful senior investigators that three years in Ann Arbor was essential to lay the foundations of a worthwhile academic career, combining our MD experience with doctorate degrees in research. NIH and Dr. Dodge obviously believed that we three trainees were worth the investment in fellowship support and time spent improving our laboratory and statistical tools for public health investigation. Nevertheless, at times, we "three blind mice" were impatient, thinking otherwise.

One morning, Dr. Dodge called me into his small conference room to meet Dr. Bocobo, a PhD in mycology working at the medical school located across the street from our School of Public Health. He was planning with Dodge how to locate the mysterious source of the Milan school yard histoplasmosis, using the latest laboratory tests in guinea pig inoculation and media culture techniques on dozens of collected soil samples.

Dr. Dodge showed us a huge map he had prepared with grid-like patterns, all numbered, to cover every square foot of soil around the school and nearby streets of Milan. A thorough master of epidemiologic detail, Dodge had researched the few successful histoplasmosis-soil investigations conducted by others. He knew of failed studies as well. He didn't like failure, not if extra effort could overcome the odds. His army field studies had prepared him well for excellence, and he was determined to pass on his hard-earned lessons to us and future trainees.

Then he had me explain to Dr. Bocobo, using the map, exactly how and where I had gathered the few soil samples (alongside the children) in the parking lot under the shade of elm trees behind and around the school last August. Both professors planned to use my small number of random soil samples as a mere pilot run in Dr. Bocobo's mycology lab to pretest his histoplasmosis isolating methodology.

SCENE THREE

Winter sets in. Ann Arbor days drag on, without sun, with the damp cold of eastern Michigan and too much indoor time in the library and the laboratory for study and exams. Attending the occasional university basketball game or ice hockey game with the children helps alleviate the grey, cabin-fever gloom of a long-endured winter season.

"Dr. Schuman," Dodge interrupts me at my study carrel. He is still formal with each of us. "Take a break, and come with me to university hospital."

"Yes, sir," I said. "What's up? You sound very fascinated."

"You may not believe this. Those random soil samples we gave Dr. Bocobo to culture in his lab were *alive* with infectious spores! Even after sitting on my office-shelf for months, waiting for Dr. Bocobo to set up his procedures."

"How do you know they were positive samples, Dr. Dodge?"

"Because, Dr. Schuman, those darn samples infected all three workers who handled them in the mycology lab! They infected Dr. Bocobo, his assistant, and one of their grad students. All three wound up with pneumonia in the hospital, confirmed by chest films and positive Histoplasmin skin tests. Of the three, poor Dr. Bocobo is most ill and feeling terrible about the accidental exposure. Even though they worked carefully under the hoods in their lab, something went terribly wrong."

I'm thinking I'm lucky that I grew up Histoplasmin positive (presumably immune) in the endemic Mississippi Valley, around St. Louis. I only hoped the two children playing beside me in the Milan school yard were equally protected by prior exposure to very low doses of spores and had also developed acquired immunity.

SCENE FOUR

Dr. Dodge and Dr. Engelke, the county health officer, were meeting with a very anxious town council in Milan. The old school library room was packed with parents of the schoolchildren. A freelance reporter and photographer from *Time* magazine was also in attendance, waiting for the epidemiologic explanation of Milan's unique school health problem. Could this happen in other nearby school districts, and nearby states, previously free of histoplasmosis?

The meticulous charts and graphs of the two skin test surveys of the elementary and high school students clearly showed a steady rate of rising H. infection beginning in kindergarten and climbing each year (each grade) with continued exposure to the mystery source. Dire news for all concerned. A hush fell over the crowded library as Dr. Dodge and the officer reported positive results from the laboratory on a dozen school ground soil samples, including my early random samples. The source of infectious spores was *definitely* the dust being churned up by the school buses in the parking lot located behind the air-intake ducts for the heating system of the school. (Dr. Bocobo's lab was shut down indefinitely for time-consuming decontamination. Further field testing in Milan was halted to protect Atlanta CDC field workers from further hazardous exposure.)

"What do you recommend, Dr. Dodge?" the chairman asked. Another hush in the room, as the professor of epidemiology took a deep breath. He had spent many an hour, day and night, pondering the most practical answer to the community's need for specific recommendations.

"Ladies and gentlemen, those lovely old elm trees around the school—your beautiful elms—are harboring flocks of starlings. You can see them morning and evening. They flock and multiply, and their droppings fertilize the soil in the parking lot. Nothing works to drive off the starlings, believe me. Others have tried, without success. They're here to stay. The soil is loaded with infectious H. spores brought in by the wind from sources along the Ohio and Mississippi valleys. The spores like the local soil, especially the bird droppings and shade conditions, and they multiply and accumulate in the dust. Chemical treatments of the dust just don't work. To protect your children, you need to do three simple practical things: one, cut down all the elms; two, pave the entire school bus parking lot with asphalt; and three, relocate the air-intake vents to a safer part of the building.

"Here is a copy of the full report from Dr. Engelke and myself for you to seriously review and, hopefully, act upon, for the good of future classes of schoolchildren."

A hush silenced the room and the crowded hallway, except for the click of the newsman's camera and its flash. Registering growing concern, the treasurer whispered something in the chairman's ear to the effect that the professor was talking about *big* money.

"Thank you, doctors," the chairman said. "The Council appreciates all your hard work on behalf of our schoolchildren. We will adjourn to digest your thirty-page report and review your recommendations, and what they might cost the taxpayers during this tight fiscal year. Good night, gentlemen."

Meanwhile, the citizens were buzzing into small group discussions in the hall. Some were arguing in the now infamous dusty parking lot under the elms.

SCENE FIVE

Three winters and springs passed. We "three blind mice," Ken, Bill, and I, completed our research theses in the required three academic years. On a beautiful but overly warm Michigan spring day in May, uncomfortably perspiring in mortar board caps and gowns, we posed in front of the university tower holding our small children while our pleased wives took photos. Our ladies, so patient and supportive, deserved the diplomas and all the credit. Soon my classmates, with their growing families, departed.

Dr. Bill was off to Colorado, Dr. Kenneth to East Lansing, and I remained in Ann Arbor. I was to help teach classes with my mentor, Dr. Dodge, repaying, in part, all that I owed him for my apprenticeship.

Milan's story was long over. A national weekly newsmagazine devoted a page to the "stricken" village, the commotion caused by histoplasmosis, and how the council finally accepted Dr. Dodge's specific recommendations. At first, to save money, they only trimmed the trees, which did nothing to discourage the roosting flocks of starlings. Later they cut down the trees and blacktopped the parking lot. Real estate values continued to slide. Eventually, with decreasing enrollment, the elementary school was closed, and all classes were moved to an annex of the secondary school building across town on the edge of the county.

The air ventilation system at the federal penitentiary was promptly and completely renovated with microfilters to protect prisoners from any future possible H. infection.

Dr. Dodge and I wondered if the successful vacuum salesman moved on to new territory, selling "spore protection" with his latest "vortex" model. The Milan story ended but not as happily as it might have. Despite an excellent epidemiologic investigation, a hallmark of Dr. Dodge's exemplary career, he had difficulty convincing the community to take action in a timely fashion, before another school year passed.

POSTSCRIPT

In 1986, while retired in California, my mentor and role model, H. J. Dodge, MD, took time to send me a handwritten letter praising my first book (*Practice-Based Epidemiology,* Gordon and

Breach Science Publishers, London). Of course, I had dedicated it to him, as well as members of my family.

Jack's warm letter recalled the "iffy" days of my first year in Ann Arbor, when I was uncertain whether I could actually make the transition from pediatric private practice to research in academe. His letter said he did not doubt my capability, if I only persevered. Evidently, our brief summer together going door-to-door in the village of Milan, tracking histoplasmosis amidst the bird droppings, actually marked the beginning of a wonderful friendship which was undimmed by passing years.

Chapter 16

Bird Flu in Our Crowded Planet: Coexisting with Other Species

SCENE ONE

Maybe it all started in Noah's Ark. The crowding, I mean, the crowding together of multiple species living together for forty days and nights during the flood. Poor Mrs. Noah had all those meals to prepare and no one but sailors to help with the housekeeping (and all that elephant-sized dung!). Maybe Noah the master carpenter had devised slanted floors in the stalls, tilted to drain the waste over the sides into the ocean—the first example of water pollution.

SCENE TWO

But now, it's 2007, and we're still crowded on our green planet Earth. Oh yes, there's a pristine, sterile space station out there to accommodate a few astronauts, but the rest of us are here on Earth are trying to survive, shoulder to shoulder, with a multitude of species, above and under the seas. After decades of research in micro-

doi:10.1300/5851_16

biology, clinical medicine, veterinary science, entomology, and ecology, we are beginning to appreciate the diversity of possible sources of human infection. We are increasingly aware of the natural exchanges of parasites between species on our crowded planet—the only home we have, so far. This is the household ark we must share with other creatures, large, small, microscopic, and even smaller.

In this millennium, we are beginning to understand the term *pandemic,* not just familiar localized garden variety, an outbreak here, an outbreak there, a spotty eruption of minor plagues among dispersed villages and scattered farmlands. Consider the history of plague crawling its way, by slow boats and slower caravans, across Medieval Europe. Nowadays infectious agents enjoy the Jet Age. People, insects, and animals with their germs, viruses and parasites can be flown in record time from continent to continent. No need to wait for history's slower transport, which might delay the transmission of infectious diseases and carriers of any kind of pathogen or parasite needing to find a new host.

So who are the players in this game of chance? This is a lottery of almost infinite possibilities for exchanges of parasitic infectious organisms—viruses, host animals, host humans, insect vectors, contaminated water and food, even fomites (articles of clothing, toothbrushes, hair combs). Don't forget the contaminated air within passenger cabins, contaminated shower heads and saunas on boats, and a huge assortment of cooking and eating utensils anywhere they are found.

SCENE THREE

I am dozing off in mid-afternoon, another sleep-deprived graduate student at the University of Michigan, trying to stay awake for a truly boring lecture with way-too-many slides. I struggle to understand the strong accent of the young scientist from New Zealand. Looking around the small conference room, half a dozen of us post-doctoral fellows are rubbing our eyes, suppressing our yawns in the semi-darkness, while the lecture slides click on and off, hypnotically.

The only one paying serious attention is our senior professor, in the front row. He's the head of the department and invited Dr. Webster all the way from his university. Dr. Thomas Francis is a seasoned MD and virologist who, decades ago, discovered the influenza B strain of virus at the Rockefeller Institute in New York. The renowned scientist's eyes gleam in the dark. Could he be remembering his own eager youthful days of discovery? This young investigator from New Zealand had some important, truly innovative observations and viral theories that will send shock waves throughout the influenza-research community, small, select and specialized as it is (London, Paris, Berlin, Hong Kong, Tel Aviv, Moscow, Tokyo, Rome, Geneva, Barcelona, Stockholm, etc.).

Only during the question and answer session, when the ceiling-lights click on, do I rouse myself, to hear the excited give-and-take between Dr. Francis and the invited guest. Dr. Webster is peppered with arguments and counterarguments; small points, large points, gaps, and possible holes in his theory are raised

Could this visitor be upsetting the prevailing apple cart of the evolution of different strains of influenza? Influenza viruses have been causing recurring human epidemics of devastating pneumonias since the early records of the "Sweating Sickness" in seventeenth century England. It was outdone by the deadly pandemic of influenza in 1917 and 1918, killing a record 20 million persons, military personnel and vastly more civilians, worldwide. That virulent pandemic had the uncanny ability to spread from person to person with the tiniest cough or sneeze on handkerchief or handshake, moist with respiratory droplets. This deadly influenza virus, with pre-antibiotic case fatality, a death rate of 30 to 50 percent, affected normally healthy, resilient youth. (Consider the fact that both of my grandfathers were afflicted in small Polish villages; both died in their forties, leaving widows and many hungry children, two of whom survived to marry in the United States and parent me, years later.)

What was young Dr. Webster's startling theory? Influenza viruses did not simply drift slowly from one human population/generation to another, from one strain (A or B with subtypes) to another human strain, in a slow five or ten year cycle, from one continent to another. By contrast, Dr. Webster believed, on the basis of a very small sample of experimental findings in his New Zealand lab, that any influenza virus strain can leapfrog, shift much more aggressively, and adapt much more virulently by passing through different animal and bird hosts, adapting from host to host.

By survival of the fittest, the very best of the viral competing strains will gain in the ability to hop from host to host, from pig to duck, from duck to chicken, from chicken to human, from human to human, gaining in its parasitic survivor power to multiply, survive in tiny droplets of secretions. Actually, the new influenza molecule does not want the host to die; like any truly successful parasite, it wants the host to live as a living, breathing, coughing incubator of the virus progeny and let the virus multiply and spread to many more hosts.

Webster's drift theory means that influenza scientists will have to leave their hospital wards and clinics, culturing sick and near-sick humans, and start to consult and work with the tiny cohort of veterinary and zoological virologists. Together the virologists must renew their efforts to monitor influenza strains among ferrets, livestock, rabbits, rodents, horses, and domestic and wild flocks of birds, especially in their travels and migrations in different geographic locales. Fortunately, the insect, reptile, and marine kingdoms are not significant players in the influenza mystery game of genetic drift, as suggested by Webster in the early 1970s.

SCENE FOUR

In a suburb of a booming Chinese metropolis, a child is dying of overwhelming pneumonia, despite the best efforts of the hospital team. Her mother is ill, but not seriously; she has a low grade fever and only coughs occasionally.

Her flock of ducks, which she raises for an occasional delicacy, are doing fine; no sign of overt avian influenza is evident. Some of her chickens, though, have been keeling over. Despite her pathetic pleas, the health officer carts away the entire flock to the overwhelmed city incinerator, where thousands of poultry are being burned around the clock, to slow the epidemic of avian influenza.

SCENE FIVE

CNN coverage of the threat of another influenza pandemic like in 1917 and 1918 is showing images of world leaders (WHO, CDC, FAO, and the UN) calling for calm, not panic, as a trickle of human cases of avian flu are being confirmed. World capitals are calling for international collaboration and sharing of anti-influenza antibiotics, medical and public personnel, available vaccine, funds for developing new DNA engineered vaccines (being pioneered by Dr. Webster and his team), and disaster strategies for coping with the possible human pandemic. How do we plan for air traffic, shipping, animal cargoes, domestic and wild bird exchanges of the virus? How will information to prepare the public be disseminated? How will emergency and hospital facilities be mobilized? Morgues and burials, for a wave of fatalities, even counter bioterrorists trying to exploit the crisis will need to be planned for. What about the closing of schools, hospitals when overfilled, or workplaces? What about disabled police, firemen, sanitation workers, and shortages of doctors and nurses? All of this will cost billions of dollars from cooperating governments, rich countries helping poor countries.

Many more questions arise for coping with the next influenza pandemic—far more questions than answers. Computer models try to predict the rate of spread and severity of illness among a population with no prior immunity, no effective vaccine, and limited efficiency of available antiviral drugs.

SCENE SIX

In the early 1880s, the economy of the southwest was being riddled by a tick infestation of their vast herds of beef cattle, the chief agricultural asset. Perfectly healthy, prime steers would leave their ranches by truck and rail from holding pens to stockyards, only to arrive sick, feverish, and unfit for the food supply. Ranchers and beef markets were going broke. An appeal for help was made to the U.S. Department of Agriculture's veterinary laboratory in Washington, DC, for some kind of cure or prevention of the devastating malady called "cattle fever."

The chief of the lab, Dr. Salmon (a veterinarian by training), assigned the lowest-ranking, youngest man on his small staff to take on the challenge. He ordered Theobald Smith (in the middle of brutal July-August heat) to work in a hard-hit area of ranches in Texas, the heartland of the cattle fever outbreak. The young doctor was, for all intents and purposes, strictly on his own.

The rest is history, heralding the birth of American bacteriology (microbiology) as a U.S. profession in its own right. No longer would the microbiologist be regarded as an ancillary lab-technician, waiting for doctors and veterinarians to call the shots. More importantly, the young scientist's intensive months of laboratory and field investigation established the critical role of the cattle's tick-cycle in carrying the protozoan (Pyroplasma) from case to case, especially under the harsh, crowded conditions in holding pens and cattle cars.

Unwittingly, perhaps because he was so young and modest and unassuming, Theobald Smith built a career in microbiology that covered decades of original research, teaching, and public service. Single-handedly he led his colleagues to raise the standards of hospital laboratory diagnosis to better serve clinicians and patients alike. His major discoveries triggered renewed interest in the role of ticks, mosquitoes, lice, and fleas as important carriers of human and veterinary illness. Smith paved the way for scientists to identify insect vectors for a host of human infections (malaria, yellow fever, sleeping sickness, plague, and typhus).

Dr. Smith and Dr. Webster are two scientists cut from the same cloth. Equally comfortable with human parasitism and animal or bird infection, equally capable of working out the critical cycles of infection and immunity, domestic hosts and wildlife, carrier states, modes of transmission, microbiological exchanges between different species, attenuation or strengthening of virulence, adaptations for insect or aerosol transmission. These men are originals, pathfinders.

Each of their contributions, and those of countless others, helps us understand this crowded planet, where even fleas have fleas, bacteria their phages, viruses their subparticles, and so on. Theobald Smith's all-embracing concept of "parasitism" as part of the cycle of coexistence of multiple species on our planet opened new horizons for many investigators. Consider how microbial guests can serve as friendly digestive bacteria in our gut or disable us with tropical fevers, including the unfriendly protozoon of malaria that can riddle one's bloodstream, one's brain, or one's liver. We should be grateful that most of our parasitic infections, bacteria, flora, and viruses are harmless, simply poachers or brief visitors to our bodies, thriving, multiplying and wending their way successfully to new hosts, without harming us.

If we are fortunate, in 2007 and 2008, avian flu virus H5N1 will migrate from ducks to migratory birds to livestock or wildlife and lose virulence for humans along the way, by means of passage through different species. That is the virologists' hope, even if our antibiotics and vaccines prove to be less than effective for influenza, this time around.

In Texas and the Midwest during the height of the cattle fever epidemic, crowding of cattle led to exchange of germs, ticks, mosquitoes, lack of exercise, feeding problems, water problems, bruising and trauma, and confinement stress. Crowding of susceptible human hosts is a key fact of life for the spread of any infection, not just from proximity but also from stress, fatigue, and lowered host resistance (immunological and otherwise). Consider crowding and the rise of epidemic crowd diseases in the early stages of the Industrial Revolution. Before that, there was the spread of plague, the rodent-flea-human cycle of the Middle Ages, which took advantage of crowded sea vessels, seaports, and emerging European cities as ripe reservoirs for the spread of contagion. Do not forget another urban-type of malady, the White Plague, person-to-person tuberculosis, which thrives on poverty, poor housing, and malnutrition.

Consider crowding of the flocks of poultry (chicken and ducks) being raised, transported, and held in countless markets in Southeast Asia, awaiting purchase and cooking for the increasingly crowded, rapidly industrializing cities in China, Thailand, Indonesia, and Vietnam.

According to the now famous virologist Dr. Webster, whose virus shift warning of inter-species exchanges went largely unheeded for twenty years by public health professionals, the simple domestic duck of China could be the Trojan horse of the 2007-2008 pandemic (if it comes to pass, Heaven forbid!).

Why the duck? Poultry scientists explain: ". . . unlike chickens, infected ducks often seem healthy, able to waddle, swim—and spread the virus in their droppings. . . . duck herders drive their flocks from one rice field to the next, following the harvest, so the birds can fatten on leftover grains" (Appenzeller, 2005).

The exchange of viruses from the herded ducks to wildlife birds ensures widespread migration across the fields and waters, across flight patterns, and, thereby, between countries, regions, and continents. It appears that I was not the only one asleep during Dr. Webster's prophetic lecture in Ann Arbor. His insight, viral observations, and prophetic theory of influenza viral shift between crowded species on our planet have been neglected by his scientific peers for too many years.

REFERENCE

Appenzeller, Tim. The Next Killer Flu. *National Geographic,* 2005 (October): 30-33.

Chapter 17

How One Small Book Changed My Career: Will Pickles in Country Practice

In this, the final chapter, I wish to conclude with an autobiographical note. In my journey to and through my medical career, I have been called many things:

- Associate editor of the Beaumont High School (St. Louis) newspaper (I never achieved Chief Editor)
- Intern and resident in pediatrics, Children's Hospital of St. Louis
- Assistant surgeon in the U.S. Public Health Service, assigned to the Mississippi State Health Department
- Private practitioner solo pediatric practice in suburban St. Louis (six years)
- Fellow in epidemiology at the University of Michigan (three years)
- Faculty, Department of Epidemiology, teaching and research, Ann Arbor, Michigan (eleven years)

In 1973, as a tenured full professor, age forty-eight, an unexpected life event changes the course of my career and our family life, as we leave Ann Arbor, Michigan to join the innovative fam-

doi:10.1300/5851_17

ily practice faculty at MUSC, in historic Charleston, South Carolina. Unlikely as it seems, one small book, a slim volume by a family practitioner in rural England, was the turning point for me.

How unlikely for one to make a transition from public health establishment type of epidemiology (the basic science for training county health officers in the United States and federal administrators in Washington, DC) to begin exploring opportunities for possible epidemiological kinds of research in a family medicine academic setting. How unlikely to move from doing large population studies of heat-wave mortality in Detroit and New York to microstudies of stress in family practice and prevention of skin cancer in farm communities.

Pulling up stakes, my wife and I moved our eight children and my ailing mother to a suburb of Charleston, not too far from beautiful Atlantic coastal beaches and tranquil sea islands. But the work in teaching and research was my challenge: How to apply the lessons of Dr. Will Pickles from 1930s and 1940s in rural England to the 1970s, 1980s and 1990s in South Carolina?

In a gem of a small book, *Epidemiology in Country Practice* (no longer in print), Dr. Pickles manages to keep meticulous records of the spread of respiratory and gastrointestinal illnesses from valley to hillside villagers, from school, from church, from the town market to the community wells, from county fairs to wedding celebrations, from one homestead farm to another, over three decades of general practice. No computers, no graduate students, no federal grants, no expert consultants. From time to time, he chats informally with college classmates who wound up as faculty scientists at Oxford or Cambridge Universities. His bedside, dispensary, and house-call records plus shoe-leather epidemiology yields discovery: the first documented outbreak of epidemic pleurisy, later found to be caused by Coxsackie B enterovirus. He calculates the incubation periods of hepatitis of two types, A and B, from his clinical epidemiological notebook of meticulous observations and his power of reasoning. He tracks the acquired immunity of crowded and less crowded households as they experience waves of communicable diseases during childhood. An amateur epidemiologist? No, a first rate observer, a pioneer in the study of family units of health and disease, in the tradition of Hippocrates.

Reading Pickles, I imagined his days as a solitary, keen observer far removed from London and Edinburgh medical centers. Reaching deep down into my character, I thought, *that's what I really want to do. But not in Ann Arbor. Maybe,* I thought, *maybe I could find "Pickles country" in a semirural state like South Carolina in the 1970s and 1980s, and beyond.*

For my first few years in Charleston, trying to define my role in a selective academic program in family medicine, I had to get used to failure. Very bright clinical residents enrolled in a three year program were falling fast asleep during my very best lectures! I experienced their lack of appreciation for my early research projects in drowning, febrile convulsions in families, life events, and stress in families ("far out, man!"). I came to learn that twenty-five-year-old residents are extremely goal-oriented, hard-nosed realists, eager to acquire the best clinical tools to use in a busy clinical practice. I tried my best to adapt to teaching them. I refocused my investigative sights from populations and subunits of the population, regional geographic studies, to the next patient you see, the next home-visit, or the next household that joins the practice. Did I succeed as a teacher-epidemiologist role model? Who knows? I know I never earned a Golden Apple Teacher Award.

I change, I grow, I have to. I have to send umpteen kids to college! I have to carve out a new academic career in my fifties. How do I adapt macro methods and strategies of epidemiological study to micro methods of smaller units of study? How do I use the medical detective approach to the family, the extended family, the rural co-op, the tomato growers on Hope Island, the pork producers of Orangeburg, the dairy farmers of Newberry, the pesticide applicators of North Charleston, the floral designers of Mt. Pleasant, and the childcare workers in Hampton County?

This became my new-old role. From Will Pickles to Stan Schuman, a very unlikely and quite amusing switcheroo! Wensleydale, England to Pickens County, South Carolina. Romantically, I could call it, going back to our mutual heritage, our beginnings, in investigational medicine and old fashioned medical detective work. Consider these heroes that inspired me, over the years:

- Jenner and the cowpox of milkmaids
- Sydenham and the London fevers
- Pasteur vaccinating sheep for anthrax
- Goldberger pioneering vitamins
- Ramazzini discovering the hazards of workers near Padua, Italy
- Moses Maimonides, rabbi and physician, showing patients and sultans how to deal with life-style stress
- Theobold Smith solving the riddle of cattle tick fever in sweltering Texas, inventing medical entomology, and expanding the concept of parasitism across species
- Jack Dodge, my Ann Arbor mentor, observing starlings and diagnosing the source of schoolyard histoplasmosis in fear-ridden Milan, Michigan, and setting a sterling example of medical detective work

CONCLUSION

When a reader is ready, a small book can make a big difference. Writing this, I wonder if this small book, my personal journal of challenges in investigative clinical epidemiology (as disease detective) can do more than amuse. Perhaps it may strike a spark somewhere, light an ember, and stimulate someone to take up this rather lonely occupation: professional observer of health and disease in the community.

Thanks, Will Pickles (wherever you are in the afterlife), for helping me find a different way, an opportunistic choice at a fork in the road—a choice that led me into intriguing episodes of exploration of the human condition as a pioneer in agromedicine.

Index

Page numbers followed by the letter "f" indicate figures and page numbers followed by the letter "e" indicate exhibits.

Rainbows in Washtubs

doi:10.1300/5851_18